This revision and classroom the new single award AQA It provides full coverage of substantive content (scient evidence) and the procedu Works, which is explained

For each sub-section of s AQA GCSE Biology speci you should be able to co knowledge and understa Many of these activities the role of science in soc our lives, and require yo develop arguments and

The activities are dealt v How Science Works pag backgrounds), which are integrated into the three main units. The points identified on these pages are designed to provide a starting point, from which you can begin to develop your own conclusions. They are not meant to be definitive or prescriptive.

At the end of each unit you will find a page of exam-style questions complete with model answers, to help you understand what is expecte the exam. Unit 1 features a combinatio choice questions and longer structured reflect the different methods of assessr and Unit 3 just have structured questio

At the end of each unit there is also a page of key words and their meanings. These pages can be used as checklists to help you with your revision. Make sure you are familiar with all the words listed and understand their meanings and relevance – they are central to your understanding of the material in that unit!

Throughout this volume, material which is higher tier only is indicated by the presence of a box with a grey background.

This book is intended as a source of first-rate revision material for GCSE students but it is our hope that it will also ease the burden of over-worked science departments.

s Book

erything you need to know in In certain places we have than the specification know to aid understanding.

de: learn actively! Constantly ooking at the text.

u think will help you to r how trivial it may seem.

hor

d as a science teacher, for 30 years before becoming r an LEA. With over 10 years' iner, she still works closely with the exam boards and has an excellent understanding of the new science specifications, which she is helping to implement in local schools.

Contents

Contents

The numbers in brackets correspond to the reference numbers on the AQA GCSE Biology specification.

How Science Works

The new AQA GCSE Biology specification incorporates two types of content:

- **Science Content** (example shown opposite)
  This is all the scientific explanations and evidence that you need to be able to recall in your exams (objective tests or written exams). It is covered on pages 12–101 of the revision guide.

- **How Science Works** (example shown opposite)
  This is a set of key concepts, relevant to all areas of science. It is concerned with how scientific evidence is obtained and the effect it has on society. More specifically, it covers…

  - the relationship between scientific evidence and scientific explanations and theories
  - the practices and procedures used to collect scientific evidence
  - the reliability and validity of scientific evidence
  - the role of science in society and the impact it has on our lives
  - how decisions are made about the use of science and technology in different situations, and the factors affecting these decisions.

Because they are interlinked, your teacher will have taught the two types of content together in your science lessons. Likewise, the questions on your exam papers are likely to combine elements from both types of content, i.e. to answer them, you will need to recall the relevant scientific facts *and* draw upon your knowledge of how science works.

The key concepts from How Science Works are summarised in this section of the revision guide. You should be familiar with all of them, especially the practices and procedures used to collect scientific data (from all your practical investigations). But make sure you work through them all. Make a note if there is anything you are unsure about and then ask your teacher for clarification.

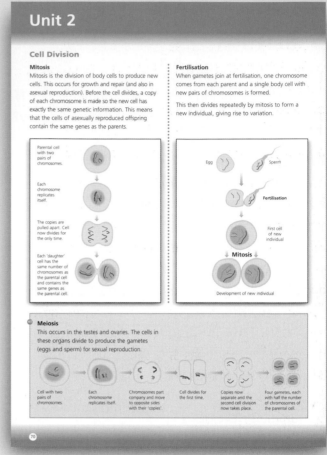

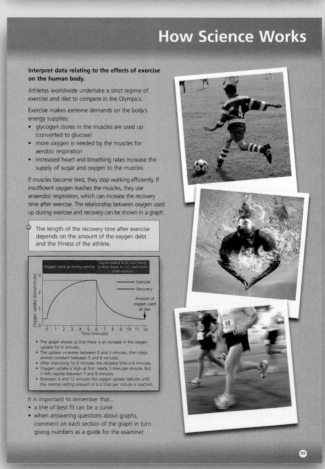

How Science Works

How to Use This Revision Guide

The AQA GCSE Biology specification includes activities for each sub-section of science content, which require you to apply your knowledge of how science works, to help develop your skills when it comes to evaluating information, developing arguments and drawing conclusions.

These activities are dealt with on the How Science Works pages (on a tinted background) throughout the revision guide. Make sure you work through them all, as questions relating to the skills, ideas and issues covered on these pages could easily come up in the exam. Bear in mind that these pages are designed to provide a starting point from which you can begin to develop your own ideas and conclusions. They are not meant to be definitive or prescriptive.

Practical tips on how to evaluate information are included in this section, on page 11.

What is the Purpose of Science?

Science attempts to explain the world we live in. The role of a scientist is to collect evidence through investigations to...

- explain phenomena (e.g. explain how and why something happens)
- solve problems.

Scientific knowledge and understanding can lead to the development of new technologies (e.g. in medicine and industry) which have a huge impact on society and the environment.

Scientific Evidence

The purpose of evidence is to provide facts which answer a specific question, and therefore support or disprove an idea or theory. In science, evidence is often based on data that has been collected by making observations and measurements.

To allow scientists to reach appropriate conclusions, evidence must be...

- **reliable**, i.e. it must be reproducible by others and therefore be trustworthy
- **valid**, i.e. it must be reliable and it must answer the question.

N.B. If data is not reliable, it cannot be valid.

To ensure scientific evidence is reliable and valid, scientists employ a range of ideas and practices which relate to...

1. **observations** – how we observe the world
2. **investigations** – designing investigations so that patterns and relationships can be identified
3. **measurements** – making measurements by selecting and using instruments effectively
4. **presenting data** – presenting and representing data
5. **conclusions** – identifying patterns and relationships and making suitable conclusions.

These five key ideas are covered in more detail on the following pages.

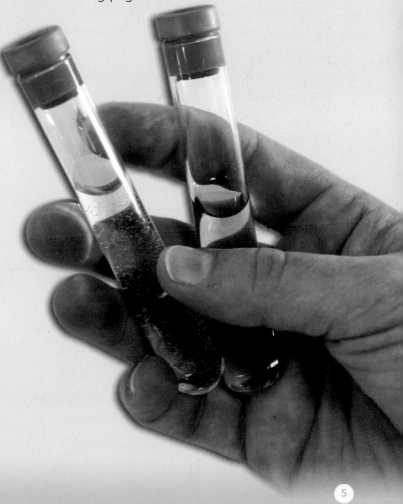

How Science Works

① Observations

Most scientific investigations begin with an observation, i.e. a scientist observes an event or phenomenon and decides to find out more about how and why it happens.

The first step is to develop a **hypothesis**, i.e. to *suggest* an explanation for the phenomenon. Hypotheses normally propose a relationship between two or more variables (factors that change). They are based on careful observations and existing scientific knowledge, and often include a bit of creative thinking.

The hypothesis is used to make a prediction, which can be tested through scientific investigation. The data collected during the investigation might support the hypothesis, show it to be untrue, or lead to the development of a new hypothesis.

Example
A biologist **observes** that freshwater shrimp are only found in certain parts of a stream.

He uses current scientific knowledge of shrimp behaviour and water flow to develop a **hypothesis**, which relates the distribution of shrimp (first variable) to the rate of water flow (second variable).

Based on this hypothesis, the biologist **predicts** that shrimp can only be found in areas of the stream where the flow rate is beneath a certain value.

The prediction is **investigated** through a survey, which looks for the presence of shrimp in different parts of the stream, representing a range of different flow rates.

The **data** shows that shrimp only occur in parts of the stream where the flow rate is below a certain value (i.e. it supports the hypothesis). However, it also shows that shrimp don't *always* occur in parts of the stream where the flow rate is below this value.

As a result, the biologist realises there must be another factor affecting the distribution of shrimp. So, he **refines his hypothesis**, to relate the distribution of shrimp (first variable) to the concentration of oxygen in the water (second variable) in parts of the stream where there is a slow flow rate.

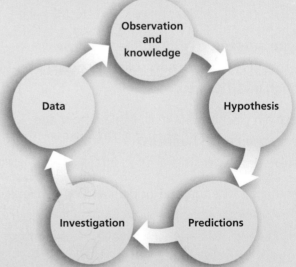

If new observations or data do not match existing explanations or theories, e.g. if unexpected behaviour is displayed, they need to be checked for reliability and validity.

In some cases it turns out that the new observations and data are valid, so existing theories and explanations have to be revised or amended. This is how scientific knowledge gradually grows and develops.

② Investigations

An investigation involves collecting data to try to determine whether there is a relationship between two variables. A variable is any factor that can take different values (i.e. change). In an investigation you have two variables:

- **independent variable**, which is controlled or known by the person carrying out the investigation. In the shrimp example on page 6, the independent variable is the flow rate of the water.
- **dependent variable**, which is measured each time a change is made to the independent variable, to see if it also changes. In the shrimp example on page 6, the dependent variable is the distribution of shrimp (i.e. whether shrimp are present or not).

Variables can have different types of values...

- **continuous variables** – can take any numerical values. These are usually measurements, e.g. temperature or height.
- **discrete variables** – can only take whole-number values. These are usually quantities, e.g. the number of shrimp in a population.
- **ordered variables** – have relative values, e.g. small, medium or large.
- **categoric variables** – have a limited number of specific values, e.g. the different breeds of dog: dalmatian, cocker spaniel, labrador etc.

Numerical values tend to be more powerful and informative than ordered variables and categoric variables. An investigation tries to establish whether an observed link between two variables is...

- **causal** – a change in one variable causes a change in the other, e.g. in a chemical reaction the rate of reaction (dependent variable) increases when the temperature of the reactants (independent variable) is increased
- **due to association** – the changes in the two variables are linked by a third variable, e.g. a link between the change in pH of a stream (first variable) and a change in the number of different species found in the stream (second variable), may be the effect of a change in the concentration of atmospheric pollutants (third variable)
- **due to chance** – the change in the two variables is unrelated; it is coincidental, e.g. in the 1940s the number of deaths due to lung cancer increased as did the amount of tar being used in road construction, however, one *did not* cause the other.

How Science Works

Fair Test

A fair test is one in which the only factor that can affect the dependent variable is the independent variable. Any other variables (outside variables) that could influence the results are kept the same.

This is a lot easier in the laboratory than in the field, where conditions (e.g. weather) cannot always be physically controlled. The impact of outside variables, like the weather, has to be reduced by ensuring all measurements are affected by the variable in the same way. For example, if you were investigating the effect of different fertilisers on the growth of tomato plants, all the plants would need to be grown in a place where they were subject to the same weather conditions.

If a survey is used to collect data, the impact of outside variables can be reduced by ensuring that the individuals in the sample are closely matched. For example, if you were investigating the effect of smoking on life expectancy, the individuals in the sample would all need to have a similar diet and lifestyle to ensure that those variables do not affect the results.

Control groups are often used in biological research. For example, in some drugs trials, a placebo (a dummy pill containing no medicine) is given to one group of volunteers – the control group – and the drug is given to another. By comparing the two groups, scientists can establish whether the drug (the independent variable) is the only variable affecting the volunteers and, therefore, whether it is a fair test.

Accuracy and Precision

In an investigation, the mean (average) of a set of repeated measurements is often calculated to overcome small variations and get a best estimate of the true value. Increasing the number of measurements taken will improve the accuracy and the reliability of their mean.

$$\text{Mean} = \frac{\text{Sum of all measurements}}{\text{Number of measurements}}$$

The purpose of an investigation will determine how accurate the data collected needs to be. For example, measures of blood alcohol levels must be accurate enough to determine whether a person is legally fit to drive.

The data collected must also be precise enough to form a valid conclusion, i.e. it should provide clear evidence for or against the hypothesis.

Fertiliser 1   Fertiliser 2   Fertiliser 3

❸ Measurements

Even if all outside variables have been controlled, there are certain factors that could still affect the reliability and validity of any measurements made:

- **the accuracy of the instruments used** – The accuracy of a measuring instrument will depend on how accurately it has been calibrated. Expensive equipment is likely to be more accurately calibrated.
- **the sensitivity of the instruments used** – The sensitivity of an instrument is determined by the smallest change in value it can detect. For example, bathroom scales are not sensitive enough to detect the changes in weight of a small baby, whereas the scales used by a midwife to monitor growth are.
- **human error** – When making measurements, random errors can occur due to a lapse in concentration and systematic (repeated) errors can occur if the instrument has not been calibrated properly or is repeatedly misused.

Any anomalous (irregular) values, e.g. values that fall well outside the range (the spread) of the other measurements, need to be examined to try to determine the cause. If they have been caused by an equipment failure or human error, it is common practice to ignore such values and discount them from any following calculations.

❹ Presenting Data

Range = Maximum value – Minimum value

Data is often presented in a format that makes the patterns more evident. This makes it easier to see the relationship between two variables. The relationship between variables can be linear (positive or negative) or directly proportional.

Clear presentation of data also makes it easier to identify any anomalous values.

The type of chart or graph used to present data will depend on the type of variable involved.

Tables organise data (patterns and anomalies in the data are not always obvious).

Height (cm)	127	165	149	147	155	161	154	138	145
Shoe size	5	8	5	6	5	5	6	4	5

Bar charts are used to display data when the independent variable is categoric or discrete and the dependent variable is continuous.

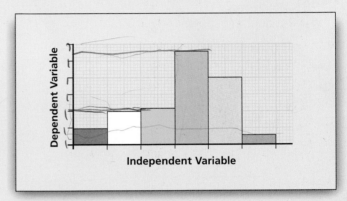

Line graphs are used to display data when both variables are continuous (see p.55).

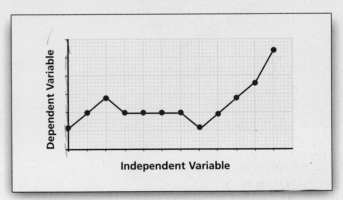

Scattergrams (or scatter diagrams) are used to show the underlying relationship between two variables. This can be made clearer by including a line of best fit.

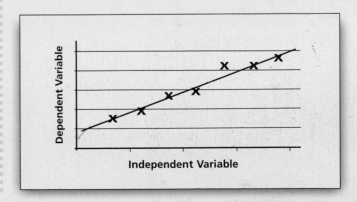

How Science Works

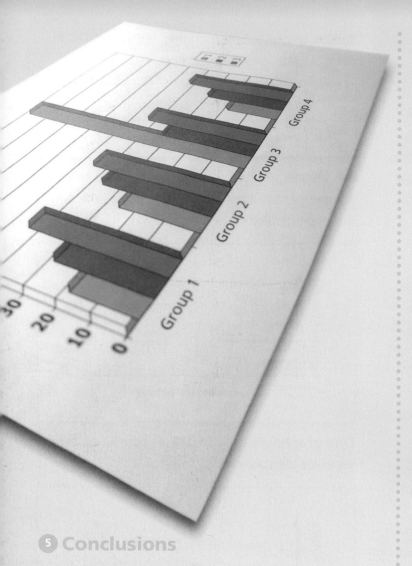

5 Conclusions

Conclusions should...
- describe the patterns and relationships between variables
- take all the data into account
- make direct reference to the original hypothesis / prediction.

Conclusions should not...
- be influenced by anything other than the data collected
- disregard any data (other than anomalous values)
- include any speculation.

Evaluation

An evaluation looks at the investigation as a whole. It should consider...
- the original purpose of the investigation
- the appropriateness of the methods and techniques used
- the reliability and validity of the data
- the validity of the conclusions (e.g. whether the original purpose was achieved).

The reliability of an investigation can be increased by...
- looking at relevant data from secondary sources
- using an alternative method to check results
- ensuring that the results can be reproduced by others.

Science and Society

Scientific understanding can lead to technological developments, which can be exploited by different groups of people for different reasons. For example, the successful development of a new drug benefits the drugs company financially and improves the quality of life for patients.

The applications of scientific and technological developments can raise certain issues. An issue is an important question that is in dispute and needs to be settled. Decisions made by individuals and society about these issues may not be based on scientific evidence alone.

Social issues are concerned with the impact on the human population of a community, city, country, or even the world.

Economic issues are concerned with money and related factors like employment and the distribution of resources. There is often an overlap between social and economic issues.

Environmental issues are concerned with the impact on the planet; its natural ecosystems and resources.

Ethical issues are concerned with what is morally right and wrong, i.e. they require a value judgement to be made about what is acceptable. As society is underpinned by a common belief system, there are certain actions that can never be justified. However, because the views of individuals are influenced by lots of different factors (e.g. faith and personal experience) there are also lots of grey areas.

Evaluating Information

It is important that you can evaluate information relating to social-scientific issues. You could be asked to do this in the exam, but it will also help you make informed decisions in life (e.g. decide whether or not to have a particular vaccination or become involved in a local recycling campaign).

When you are asked to **evaluate** information, start by making a list of the pluses and the minuses. Then work through the two lists, and for each point consider how this might impact on society. Remember, **PMI** – pluses, minuses, impact on society.

You also need to be sure that the source of information is reliable and credible. Here are some important factors to consider:

- **opinion**
  Opinions are personal viewpoints. Opinions which are backed up by valid and reliable evidence carry far more weight than those based on non-scientific ideas (e.g. hearsay or urban myths).

- **bias**
  Information is biased if it does not provide a balanced account; it favours a particular viewpoint. Biased information might include incomplete evidence or try to influence how you interpret the evidence. For example, a drugs company might highlight the benefits of their drugs but downplay the side effects in order to increase sales.

- **weight of evidence**
  Scientific evidence can be given undue weight or dismissed too lightly due to...
  - political significance, e.g. evidence that is likely to provoke an extreme and negative reaction from the public might be downplayed
  - status (academic or professional status, experience, authority and reputation), e.g. evidence is likely to be given more weight if it comes from someone who is a recognised expert in that particular field.

Limitations of Science

Science can help us in lots of ways but it cannot supply all the answers. We are still finding out about things and developing our scientific knowledge. There are some questions that we cannot answer, maybe because we do not have enough reliable and valid evidence.

There are some questions that science cannot answer at all. These tend to be questions relating to ethical issues, where beliefs and opinions are important, or to situations where we cannot collect reliable and valid scientific evidence. In other words, science can often tell us whether something *can* be done and *how* it can be done, but it cannot tell us whether it *should* be done.

Unit 1

How do human bodies respond to changes inside them and to their environment?

The nervous system and hormones allow the human body to respond to changes. To understand this, you need to know...
- what the nervous system and hormones are
- how the brain coordinates a response
- how hormones regulate functions and how they are involved in controlling fertility.

Parts of the Nervous System

The nervous system consists of the **brain**, the **spinal cord**, the **spinal nerves** and **receptors**. It allows organisms to react to their surroundings and to coordinate their behaviour. Information from **receptors** passes along **neurones** (nerve cells) to the brain which coordinates the response.

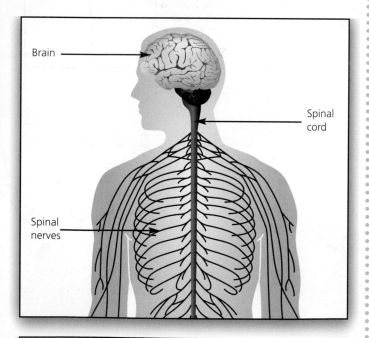

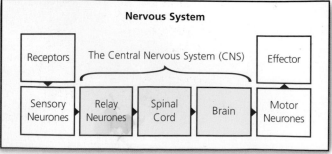

Nervous System

The Three Types of Neurone

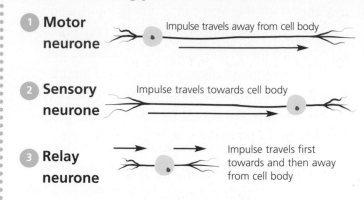

1. **Motor neurone** — Impulse travels away from cell body

2. **Sensory neurone** — Impulse travels towards cell body

3. **Relay neurone** — Impulse travels first towards and then away from cell body

Neurones are specially adapted cells that can carry an electrical signal, e.g. a nerve impulse.

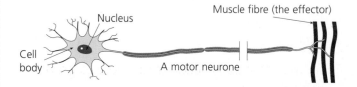

Cell body / Nucleus / A motor neurone / Muscle fibre (the effector)

They are elongated (stretched out) to make connections between parts of the body. They have branched endings which allow a single neurone to act on many muscle fibres. The cell body has many connections to allow communication with other neurones.

Connections Between Neurones

Neurones do not touch each other; there is a very small gap between them called a **synapse** (see diagram below). When an electrical impulse reaches the gap via Neurone A, a chemical transmitter is released, which activates receptors on Neurone B and causes an electrical impulse to be generated in Neurone B. The chemical transmitter is then destroyed.

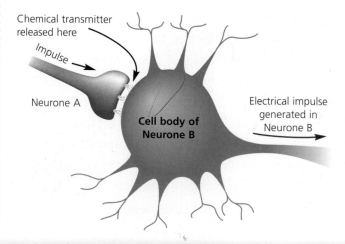

Chemical transmitter released here / Impulse / Neurone A / **Cell body of Neurone B** / Electrical impulse generated in Neurone B

Types of Receptor

Receptors detect (respond to) stimuli which include light, sound, changes in position, tastes, smells, touch, pressure, pain and temperature.

- **Light** – receptors in the eyes.
- **Sound** – receptors in the ears.
- **Change of position** – receptors in the ears (balance).
- **Taste** – receptors on the tongue.
- **Smell** – receptors in the nose.
- **Touch, pressure, pain and temperature** – receptors in the skin.

Conscious Action

The pathway for receiving information and then acting upon it is as follows:

Stimulus	Receptors	Coordinator	Effectors	Response
Loud music	Sound-sensitive receptors in the ear	**Sensory Neurones** ▼ **Central Nervous System** ▼ **Motor Neurones**	Muscles in arms and fingers	Turn music down

The coordinator is the central nervous system, to which impulses are transmitted via the spinal nerves.

Reflex Action

Sometimes conscious action is too slow to prevent harm to the body, e.g. removing your hand from a hot plate. **Reflex action** speeds up the response time by missing out the brain completely. The spinal cord acts as the coordinator and passes impulses directly from a sensory neurone to a motor neurone via a relay neurone, which by-passes the brain. Reflex actions are automatic and quick.

Stimulus	Receptors	Coordinator	Effectors	Response
Drawing pin	Pain receptor	**Sensory Neurones** ▼ **Relay Neurone in Spinal Cord** ▼ **Motor Neurones**	Muscles to hand	Withdraw hand

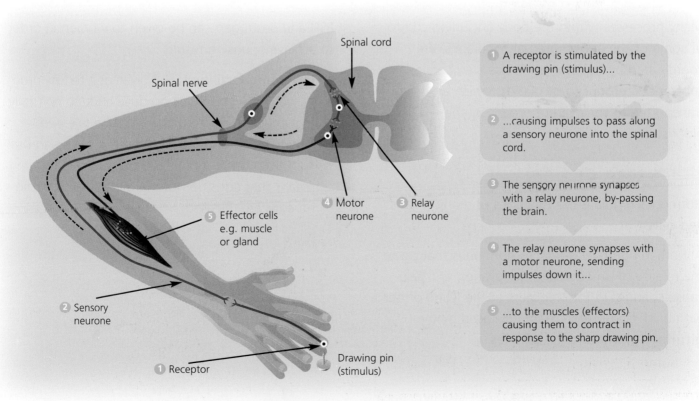

Spinal cord

Spinal nerve

4 Motor neurone

3 Relay neurone

5 Effector cells e.g. muscle or gland

2 Sensory neurone

1 Receptor

Drawing pin (stimulus)

1 A receptor is stimulated by the drawing pin (stimulus)...

2 ...causing impulses to pass along a sensory neurone into the spinal cord.

3 The sensory neurone synapses with a relay neurone, by-passing the brain.

4 The relay neurone synapses with a motor neurone, sending impulses down it...

5 ...to the muscles (effectors) causing them to contract in response to the sharp drawing pin.

Unit 1

Internal Conditions

Humans need to keep their internal environment relatively constant. Body temperature, and the levels of water, salts (ions) and blood sugar need to be carefully controlled.

- **Temperature** (ideally 37°C – the temperature at which most body enzymes work best.) Temperature is increased by shivering and narrowing skin capillaries. Temperature is decreased by sweating and expanding skin capillaries.
- **Water content**
  Water is gained by drinking, and lost by breathing via the lungs and sweating. Any excess is lost via the kidneys in urine.
- **Ion content** (sodium, potassium, etc.)
  Ions are gained by eating and drinking. Ions are lost by sweating and excess is lost via the kidneys in urine.
- **Blood sugar (glucose) levels**
  Glucose provides the cells with a constant supply of energy. It is gained by eating and drinking.

How Conditions are Controlled

Many processes within the body (including control of some of the above internal conditions) are coordinated by **hormones**. These are chemical substances, produced by glands, which are transported to their target organs by the bloodstream.

Hormones and Fertility

Hormones regulate the functions of many organs and cells.

A woman naturally produces hormones that cause the release of an egg from her ovaries, and also cause changes in the thickness of the lining of her womb. These hormones are produced by the **pituitary gland** and the **ovaries**.

Natural Control of Fertility

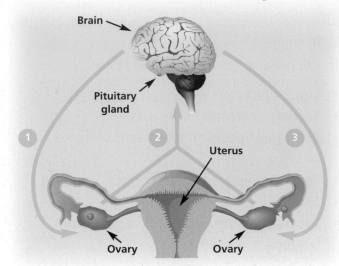

1. **Follicle stimulating hormone (FSH)** from the pituitary gland causes the ovaries to produce oestrogen and an egg to mature.
2. **Oestrogen**, produced in the ovaries, inhibits the production of FSH and causes the production of luteinising hormone (LH).
3. **LH**, also from the pituitary gland, stimulates the release of an egg in the middle of the menstrual cycle.

Artificial Control of Fertility

FSH and oestrogen can be given to women in order to achieve opposing results.

- **Increasing fertility:** FSH is given as a fertility drug to women who do not produce enough naturally, to stimulate eggs to mature and be released.
- **Reducing fertility:** oestrogen is given as an oral contraceptive to inhibit FSH production. This means that eggs do not mature in the ovary, so no eggs are released.

You need to be able to evaluate the claims made by a manufacturer about a sports drink.

Example

SustainEnergy

This carbohydrate drink will help to increase your available energy, and improve hydration and performance when taken before and during exercise. You will develop endurance, and be able to exercise more intensely and for longer. It will also help you to recover more rapidly so that you can train more frequently.

SustainEnergy. For all your energy needs.

– High in glucose
– Rehydrates (contains water)
– Matches body ions
– No artificial additives

Benefits

- Contains glucose to provide cells with a constant supply of energy.
- Improves hydration by replacing water lost from the body via lungs when we breathe out and via skin when we sweat during exercise.
- Replaces ions lost through skin when we sweat.
- SustainEnergy should have a positive impact on the fitness of those who drink it.

Problems

- Encourages over-exercise before the body has chance to recover.
- No evidence that its claims are true.
- No reference to which of the body's ions are present.
- The claims could be influenced by the manufacturer's need to make a financial gain from this product.

It would be necessary to investigate this product further and to compare it with others on the market before arriving at a final conclusion.

How Science Works

You need to be able to evaluate the benefits of, and problems that may arise from, IVF treatment, including its impact on society.

Example

Issue 19
January 2006

SCIE

Natural Conception

Did you know that one in seven couples in the UK experiences delays in conceiving? Only a third of those trying for a baby become pregnant each month, although some of these will miscarry before they even know they are pregnant. The most common reasons for miscarriage / fertility problems include hormonal problems and blocked fallopian tubes in women, and a low sperm count in men.

What can Specialist Scientists do?

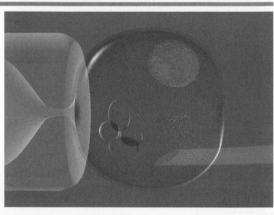

Once a couple has been referred to a fertility specialist (called a reproductive medicine specialist), the treatment they receive will depend on the cause of the problem and can range from hormone treatments such as FSH, which boosts egg production, to assisted conception techniques such as *in vitro* fertilisation (IVF).

In Vitro Fertilisation

IVF involves removing eggs from the woman's ovaries and mixing them with her partner's sperm or donated sperm in a laboratory. A number of eggs are then placed back in the womb. This is a demanding treatment for the woman and the success rate is just 15 per cent. Many cycles of treatment may be required.

Although access to IVF is limited on the NHS, couples are now entitled to one cycle of treatment. To qualify, women must be between 23 and 39 and have a specific fertility problem (e.g. blocked fallopian tubes) or have failed to conceive for three years despite regular intercourse. A cycle of IVF at a private clinic costs around £2,000.

Benefits	Problems
• Can help a woman become pregnant. • Uses woman's own eggs and partner's own sperm. • Provides an alternative to adoption.	• Emotionally / physically demanding treatment. • Only 15% success rate. • Age restriction. • Costly. • Could lead to over population. • Uses NHS resources. • Increases expectation for babies on demand.

11.2

What can we do to keep our bodies healthy?

A nutritious, balanced diet and regular exercise are required to keep our bodies healthy. To understand this, you need to know...

- what a healthy diet consists of
- what can happen as a result of a poor diet
- how the metabolic rate varies
- how to balance cholesterols to maintain a healthy heart
- what the problems are with fats, salt and processed food.

A combination of a balanced diet and regular exercise is needed to keep our bodies healthy.

Metabolic Rate

This is the rate at which all the chemical reactions in the cells of the body are carried out. It varies with...

- the amount of activity you do – your metabolic rate increases with the amount of exercise you do and it stays high for some time after you have finished exercising
- the proportion of fat to muscle in your body
- your family history – it can be affected by inherited factors.

People who exercise regularly are usually fitter than those who do not.

The less exercise you take and the warmer it is, the less food you need.

Healthy Diets

A healthy diet contains the correct balance of the different foods your body needs:

- carbohydrates
- fats
- protein
- fibre
- vitamins
- minerals
- water.

The proportions required depend on an individual's body and lifestyle.

A person is **malnourished** if their diet is not balanced. A poor diet can lead to...

- a person being too fat or too thin
- deficiency diseases, for example, scurvy, which is caused by lack of vitamin C.

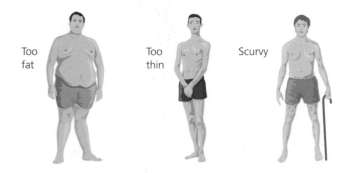

Too fat   Too thin   Scurvy

Health Problems Linked to Diet and Exercise

In the **developed** world (e.g. UK, USA) people are consuming too much food and taking too little exercise. The result is high levels of obesity and diseases linked to excess weight, such as...

- arthritis (worn joints)
- diabetes (high blood sugar)
- high blood pressure
- heart disease.

In the **developing** world (e.g. some parts of Africa and Asia) people suffer from problems linked to lack of food, such as...

- reduced resistance to infection
- irregular periods in women.

Unit 1

How to Improve Our Diet

Two things we can control to improve our diets are...

- cholesterol
- salt.

Cholesterol

Cholesterol is naturally made in the liver and is found in the blood. Diet and inheritance factors affect how much cholesterol the liver makes.

Cholesterol is carried around the body by chemicals called **lipoproteins** (made from fat / lipid and protein). High levels of low-density cholesterol (bad cholesterol) in the blood increase the risk of disease of the heart and blood vessels.

- **Low-density lipoproteins** (LDLs) can cause heart disease.
- **High-density lipoproteins** (HDLs) are good cholesterols.

The balance of these is important for a healthy heart.

Salt

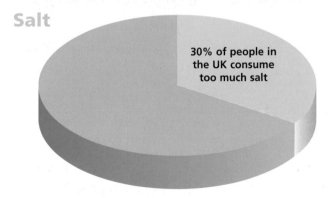

30% of people in the UK consume too much salt

Too much salt in the diet can lead to increased blood pressure. So you should...

- eat less **salt**
- eat less **processed food**, which often contains a high proportion of fat and / or salt
- eat less **saturated fats** e.g. butter, animal fat (these increase blood cholesterol)
- eat more **unsaturated fats** (monounsaturated and polyunsaturated fat) e.g. vegetable oil, some fish. These help to reduce blood cholesterol levels and improve the balance between good and bad lipoproteins.

Ideal amounts of cholesterol in the blood

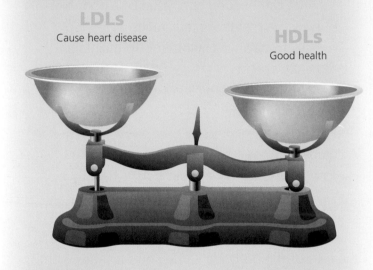

LDLs
Cause heart disease

HDLs
Good health

More...

Less...

You need to be able to evaluate information about the effect of food on health.

Example

To be healthy, the main thing is to ensure you have a balanced intake of protein, carbohydrates, essential fats, fruit, fibre, vitamins and minerals.

The Western Approach to a healthy diet involves limiting your intake of salt, alcohol, caffeine and saturated fats, and eating at least five portions of fruit and vegetables per day. People should eat plenty of fresh and wholegrain foods and have only a moderate amount of sugar.

Benefits	Problems
• Balances protein, carbohydrates and vitamins daily throughout the year. • Foods are available all year round. • Encourages eating fruit and vegetables every day.	• Some people think that the body reacts best to foods that are seasonal and eaten at the time of the year when they are naturally available. • In order for certain foods to be available all year round, chemical / genetic assistance may be used to grow foods out of season.

The Oriental Approach involves eating foods in the season they are naturally grown, so foods such as salads, soft fruits, peas and beans are eaten in summer, whereas in winter, vegetables grown below ground, like carrots and turnips, are eaten.

All foods are categorised as being either 'warming' or 'cooling'. Warming foods include meats, pulses and ginger, and are boiled, grilled or fried. Cooling foods include lettuce, watercress and soft fruits that are eaten raw, cool or refrigerated.

Benefits	Problems
• Some people think this is the healthiest option because it involves eating seasonal food. • Encourages consumption of healthy, natural foods such as salad and fruit. • Foods are not genetically modified.	• The diet keeps changing which can upset some people's stomach. • During certain times of the year, e.g. winter, the diet may be restrictive.

How Science Works

You need to be able to evaluate claims made by a slimming programme.

Example

April 2006

Slimm

Slimmer April 2006

Advertorial

The all-new <u>low-carb</u> diet

It's easy to lose weight quickly on this new slimming programme. All you have to do is limit your carbohydrate intake.

You can consume a very small amount of carbohydrates from vegetables such as broccoli, but you are not allowed any carbohydrates from bread, pasta or fruit.

- Eat unlimited quantities of protein and fats (e.g. meat, cheese, eggs)!
- No need to make yourself eat fruit!
- No need for exercise!

How does it work?

Managing the intake of carbohydrates is certain to lead to weight loss because without carbohydrates to provide energy, the body starts using up stored fat as the main source of energy.

Page 3

Benefits	Problems
• Low-carbohydrate diets can help you to lose weight very quickly in the short term. • Low-carbohydrate diets can seem like an easy way to lose weight. • You can still eat many of the foods that would be cut out on a conventional slimmer's diet, e.g. burgers, cheese.	• Such a drastic change in diet can have a negative effect on the body. • Exercise is not encouraged by this slimming programme, but it is required for a person to be healthy – especially when they are consuming so much fat. • The body needs a healthy balance of foods to maintain good health: low-carbohydrate diets are not balanced and not good in the long term. • The body misses out on essential nutrients found in fruit, vegetables and grains. • Low-carbohydrate diets do not allow sufficient consumption of fruit and vegetables to meet the recommended daily allowance. • Long-term high intake of protein puts a strain on the kidneys. • Lack of energy (energy should come from carbohydrates). • In the absence of carbohydrates, protein (muscle) stores get used for energy, as well as fat.

11.3

How do we use / abuse medical and recreational drugs?

Drugs are used in medicine to cure illnesses and diseases. Drugs, alcohol and tobacco, may also be used recreationally by people who like the effect they have. To understand this, you need to know about...

- the benefits and the harmful effects of drugs
- the stages in the development and testing of drugs
- thalidomide and its effects
- the effects of legal and illegal drugs
- the effects of tobacco and alcohol on the human body.

Drugs

Drugs are chemical substances which alter the way the body works. They can be beneficial but may also harm the body. Some drugs can be obtained from natural substances (many have been known to indigenous people for years). Others are synthetic (man-made) and need to be thoroughly tested and trialled in the laboratory to find out if they are toxic (poisonous). Then they are checked for side-effects on human volunteers.

Developing New Drugs

The flow chart below shows the stages in developing new drugs.

New drug made in a laboratory

↓

Tested in laboratory for toxicity

↓

Trialled on volunteers to check for side-effects

Example: Thalidomide

- This drug was developed as a sleeping pill. It was tested and **approved** for this use.
- It was also found to be effective in relieving morning sickness in pregnant women. It had **not been tested** for this use.
- Many babies born to mothers who took the drug had severe limb abnormalities, so it was **banned**.
- Thalidomide was re-tested and is now used successfully to treat leprosy.

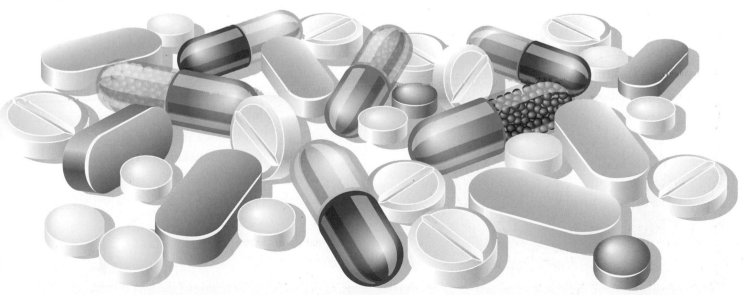

Unit 1

Legal and Illegal Drugs

Some drugs are used illegally for pleasure (recreation) but they can be very harmful.

Drugs alter chemical processes in people so they may become dependent or addicted to them. They may therefore suffer withdrawal symptoms if they do not have them. Symptoms may be psychological or physical (e.g. paranoia, sweating, vomiting).

Heroin and cocaine are examples of two very addictive drugs, which are used illegally.

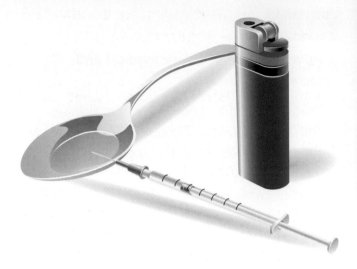

Alcohol	Effects
Contains the chemical ethanol. 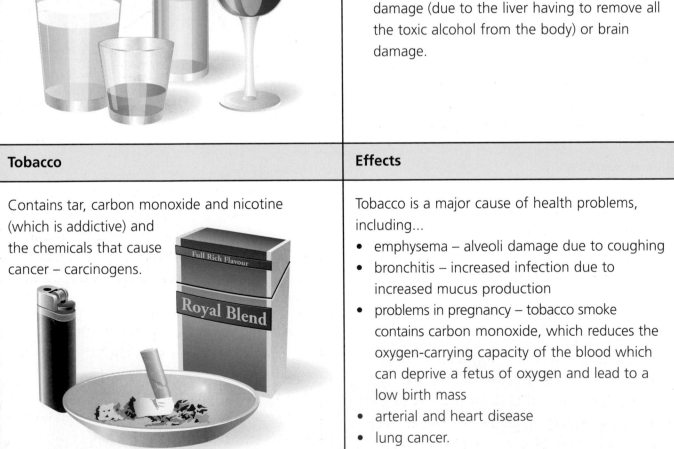	• Alcohol is a depressant – it causes reactions to slow down. • Alcohol can lead to a lack of self-control. • Excess alcohol can lead to unconsciousness and even coma or death. • The long-term effects of alcohol can be liver damage (due to the liver having to remove all the toxic alcohol from the body) or brain damage.
Tobacco	**Effects**
Contains tar, carbon monoxide and nicotine (which is addictive) and the chemicals that cause cancer – carcinogens. Full Rich Flavour Royal Blend	Tobacco is a major cause of health problems, including... • emphysema – alveoli damage due to coughing • bronchitis – increased infection due to increased mucus production • problems in pregnancy – tobacco smoke contains carbon monoxide, which reduces the oxygen-carrying capacity of the blood which can deprive a fetus of oxygen and lead to a low birth mass • arterial and heart disease • lung cancer.

You need to be able to evaluate the effect of statins on cardio-vascular disease.

Example

– *Why does cholesterol need to be reduced?*

High levels of cholesterol, especially the 'bad' type (LDLs), can cause the arteries to clog up with a thick fatty substance. The result is the narrowing of the arteries that take blood to heart muscle (coronary arteries). This narrowing and hardening of the arteries is called atherosclerosis, and can cause heart attacks. A sufferer will feel a heavy, tight chest pain and perhaps experience breathing difficulties. In very serious cases, a coronary artery may become blocked by a blood clot (thrombosis). This can cause severe pain and is life-threatening.

– *What are statins?*

Statins are drugs that work in the liver to reduce the manufacture of cholesterol. They have been found to be the most effective drugs for lowering LDL levels.

– *How long do you take them for?*

Statins are taken for life. They can reduce the chance of having a heart attack or a stroke by up to a third, and can increase the life expectancy of a person with a history of high cholesterol when taken long term.

– *Are there any side effects?*

They may cause headaches, sickness, diarrhoea, insomnia, liver problems, stomach upsets, hepatitis and muscle aches. They cannot be taken by children, pregnant women or heavy drinkers.

– *Is there an alternative?*

If your doctor prescribes statins, you should take them. However, you can reduce the need for them if you stop smoking, eat a healthy diet with plenty of fruit and vegetables and low fat and salt content, and exercise regularly: moderate exercise three times a week reduces the risk of a heart attack by a third.

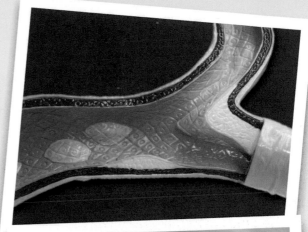

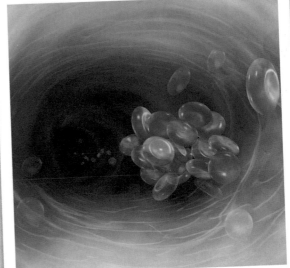

Above: Red blood cells travelling down an artery.
Top: Artery 'Y' section with cutaway portions shows build-up of cholesterol deposits in several locations.

Benefits
• They reduce the manufacture of cholesterol.
• They reduce the risk of heart attack or stroke caused by cardio-vascular disease by up to a third.
• They are the most effective drugs for reducing LDLs.

Problems
• Statins can have many side effects.
• They cannot be taken by children, pregnant women or heavy drinkers.
• Availability of statins reduces the need for people to change their unhealthy lifestyles.
• They must be taken for life.

How Science Works

You need to be able to evaluate the different types of drugs and why some people use illegal drugs for recreation.

Example
The table below shows the different types of illegal drugs commonly used in the UK. Drugs are classified into class A, B and C; class A is the most dangerous.

Drug	Class	Details of Drug	Number of people using it regularly
Heroin	A	Sedative. Smoked or injected. Causes severe cravings, very addictive.	43 000
Cocaine	A	Stimulant, increases confidence, raises heart rate and blood pressure. Injected, inhaled or smoked. Causes cravings, very addictive.	755 000
Ecstacy	A	Stimulant, gives adrenaline rush, feeling of well-being, high body temperature, anxiety. Ingested in tablet form.	614 000
Amphetamines	A / B	Stimulant, increases heart rate. Ingested in tablet form. Can be very addictive and can cause paranoia.	483 000
Cannabis	C	Relaxant. Smoked. Can cause hallucinations.	3 364 000

Types of Drug Use	Problems
Experimental: when people try a drug for the first few times out of curiosity, boredom, or because their friends are doing it. Some do it to rebel against authority.	• A person is just as likely to have a bad reaction from trying a drug the first time as a person who uses drugs for recreation regularly. • It could lead to recreational use. • It could lead to trying harder drugs to get a new 'high'.
Recreational: when people use drugs in a regular but fairly controlled way. It is seen by some people as a way of relaxing.	• It could lead to dependent drug use. • It could lead to using harder drugs to get a new 'high'. • A person risks a bad reaction to drugs every time they use them. • It could lead to mental problems.
Dependent: when people are addicted to drugs and are dependent on the feeling it gives them. It is also physically addictive – the body needs the drug.	• People can not function without the drug. • People suffer withdrawal symptoms if they do not have the drug. • A person risks a bad reaction to drugs every time they use them. • It could lead to mental problems.

Drugs are illegal so there are no guarantees that you are getting what you think you are getting and they are very expensive.

In all cases, a harmful substance is being put into the body, so there is always a risk of death.

How Science Works

You need to be able to evaluate claims made about the effect of cannabis on health and the link between cannabis and addiction to hard drugs.

Example

New concerns over cannabis use

A leading report published yesterday has sparked a fresh wave of concern about the dangers of using cannabis.

Concern about cannabis use is not isolated to the 21st century.

In Egypt in the 8th Century, laws were introduced which prohibited the use of 'hemp drugs' (cannabis). In the 19th Century, a large scale investigation into the health effects (physical and mental) of cannabis use was set up by the Indian Hemp Drugs Commission who concluded that the link between cannabis and mental injury was complex. In recent years, research in Scandinavia has linked cannabis to severe mental illness.

The new report backs up these claims that it is bad for your health, and suggests that in many cases users go on to use harder drugs, which can be extremely addictive.

— 1 —

The report claims that cannabis contains more tar and more cancer-causing chemicals than cigarettes: around 400 chemicals which are known to affect the brain.

It claims that its use affects blood pressure, causing fainting, or more severe problems for people with heart and circulation conditions. Although cannabis is not as addictive as alcohol, tobacco or amphetamines, many users do become psychologically addicted, which may then lead to them trying harder drugs.

Other reports, however, have said that cannabis, unlike harder drugs, does not lead to major health problems, and few deny that it can have beneficial effects for patients suffering from conditions such as HIV, multiple sclerosis and cancer.

The recreational use of cannabis is still illegal in Britain although up to half of all young people have tried it.

— 2 —

Is smoking cannabis bad for your health?

Yes

- Cannabis has a higher concentration of cancer-causing substances (carcinogens) than tobacco.
- It has a higher tar content than tobacco so can lead to bronchitis, emphysema and lung cancer.
- It can increase the risk of fainting since it disrupts the control of blood pressure.
- It can lead to mental illness.
- It contains more than 400 chemicals. The main one that affects the brain is known as THC.
- People with heart and circulation disorders or mental illness can be adversely affected by it.
- It may be psychologically addictive.

No

- Its effects are beneficial to patients suffering from various medical conditions including HIV, multiple sclerosis and cancer.
- Unlike harder drugs, a government report has suggested that high use of cannabis is not associated with major health or sociological problems.
- Cannabis is less addictive than amphetamines, tobacco or alcohol, and does less harm to the body.

Can smoking cannabis lead users to harder drugs?

Yes

- Cannabis may be a 'gateway' drug to more addictive and harmful substances such as heroin and cocaine.

No

- Many cannabis smokers never use any harder drugs.

How Science Works

You need to be able to explain how the link between smoking tobacco and lung cancer gradually became accepted.

During the 1940s and 1950s, there was a marked increase in deaths from lung cancer which prompted scientists to investigate the cause.

1940–1950 – Many scientists thought pollution was to blame for the increase in lung cancer deaths, whilst others believed the cause was the tar used in new roads.

Sir Richard Doll was commissioned by the Medical Research Council to investigate a possible link between smoking tobacco and lung cancer. He visited 2000 people suspected of having lung cancer across England and found that those who had the disease were heavy smokers. Those who did not have it did not smoke. However, the findings, published in 1950, were widely ignored.

1954 – The government accepted that there was a strong link between smoking tobacco and lung cancer.

1970s – In the 1970s the link began to be taken more seriously, and the media began to discuss it openly.

1990s – Some people in the tobacco industry continued to dispute claims of the link until as late as 1997, when they were sued by 40 US states to pay for the treatment costs of tobacco-related illnesses, and were forced to accept the evidence.

2000s – It is now accepted that people who smoke have an increased risk of lung cancer; the more people smoke, the more the risk is increased. The number of smokers in England has fallen (30% of men smoked in 2000 compared to 80% in 1950).

2006 – The UK Government passed a bill to ban smoking in all enclosed public places. It is already law in Scotland and will extend to England and Northern Ireland in 2007.

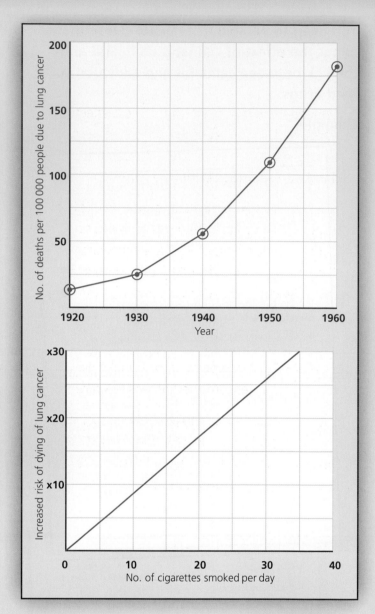

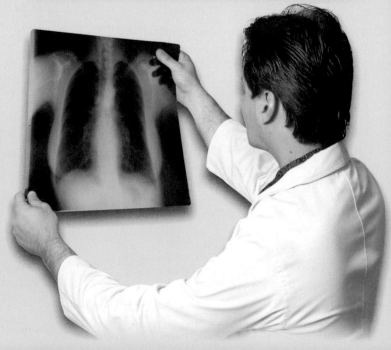

You need to be able to evaluate different ways of trying to stop smoking.

Example

A test was carried out to find the most effective way to stop smoking. The four methods tested were: nicotine patches, acupuncture, hypnosis and 'going cold turkey' (the **independent variables**).

400 participants who wanted to quit were divided into four equal groups. Each group tried one of the four methods for 2 months. The effectiveness of each method was measured by the number of people in each group who were still not smoking one year later (the **dependent variable**).

Because the study involved a large number of participants, it was important that similar people were selected to ensure the findings were **reliable**.

Therefore, people who smoked between 20 and 25 cigarettes a day were selected, and divided into groups so there was an equal number of men and women in each, **limiting the variables**.

Here are the results of the investigation.

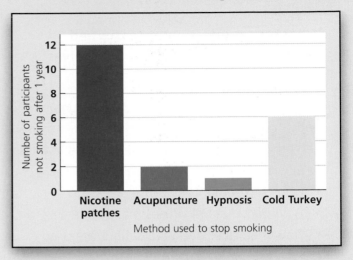

The bar graph shows us that nicotine patches are the most effective way to stop smoking. They are twice as effective as willpower alone (cold turkey).

Acupuncture and hypnosis are significantly less effective.

The conclusion is limited by the data and further tests would be needed to see if this were true for the whole population.

Method	Benefits	Problems
Nicotine patches (oversize sticking plasters containing nicotine).	• Fairly cheap. • Does not require expert help. • Easier than using willpower alone.	• Continues to put chemicals into the body. • People can become reliant on patches. • Unsuitable to use during pregnancy or if you have heart problems. • Can cause skin irritation or allergies.
Acupuncture (fine needles inserted in specific areas of the skin).	• Helps to restore the body's natural balance of health.	• Not proven to be very effective. • Must be administered by an expert. • Expensive.
Hypnosis (trained hypnotherapist suggests stopping while person is in a relaxed state).	• No side effects.	• No scientific evidence to support this. • Expensive. • Must be administered by an expert. • Not proven to be very effective.
Cold turkey (willpower only).	• No drugs or chemicals are used. • Quick. • No cost involved.	• Difficult to rely on willpower alone. • May experience withdrawal symptoms, e.g. mood swings, nausea.

Unit 1

11.4

What causes infectious diseases and how can our bodies defend themselves against them?

Microorganisms can thrive inside the human body, so the body has to try to stop them from getting in, and defend itself against those that do get in. To understand this, you need to know…

- how microorganisms cause disease
- how bacteria and viruses make us feel ill
- how medicines and vaccinations are used
- how immunisation works.

Bacteria and Viruses

Microorganisms that cause infectious diseases are called **pathogens**. Bacteria and viruses are the two main types of pathogen which may affect health.

Bacteria	Viruses
Very small.	Even smaller.
Reproduce very quickly.	Reproduce very quickly once inside living cells, which are then damaged.
Can produce toxins (poisons) which make us feel ill.	Can produce toxins (poisons) which make us feel ill.
Responsible for illnesses like tetanus, cholera, tuberculosis.	Responsible for illnesses such as colds, flu, measles, polio.

Defence Against Pathogens

White blood cells form part of the body's **immune system** and help to fight infection by ingesting pathogens, producing antitoxins to neutralise toxins produced by the pathogens, and producing **antibodies** to destroy particular pathogens.

Treatment of Disease

The symptoms of disease are often alleviated using painkillers. You will be familiar with the mild versions of these, e.g. aspirin. Although painkillers are useful, they do not kill pathogens.

Antibiotics like penicillin are often used against bacteria. They kill infective bacteria inside the body, but cannot be used to kill viruses which live and reproduce inside cells. It is difficult to develop drugs which kill viruses without damaging the body's tissues.

Over-use of Antibiotics

Many strains of bacteria, including MRSA, have developed resistance to antibiotics as a result of natural selection.

Some individual bacteria in a particular strain have natural resistance. If the majority of the strain is wiped out by antibiotics, this leaves the field clear for the resistant bacteria to multiply quickly, passing on their resistance. It is therefore necessary to have a range of antibiotics and to select the one that is most effective for treatment of a particular infection.

Development of further resistance is avoided by preventing over-use of antibiotics.

Vaccination

A person can acquire **immunity** to a particular disease by being vaccinated (immunised).

1. An inactive / dead pathogen is injected into the body.
2. White blood cells produce antibodies to destroy the pathogen.
3. The body then has an acquired immunity to this particular pathogen since the white blood cells are sensitised to it and will respond to any future infection by producing antibodies quickly. An example is the MMR vaccine used to protect children against measles, mumps and rubella.

You need to be able to relate the contribution of Semmelweiss in controlling infection to solving modern problems with the spread of infection in hospitals.

The work of Semmelweiss, and subsequent scientists, has led to the creation of many regulations which help to maintain hygiene standards in modern hospitals and reduce the chance of infections being spread.

Example

1865

Semmelweiss dies of blood poisoning

The Hungarian doctor, Ignaz Semmelweiss, died today after contracting blood poisoning.

He will be remembered for his work in local hospitals where he reduced patient deaths on his wards from 12% to 1% by insisting that doctors washed their hands after surgery and before visiting another patient.

He had recognised that germs on surgeons' and doctors' hands were infectious and contagious and were responsible for many patients' deaths.

Frampton General Hospital

Health and Safety Regulations

- All staff must wash their hands thoroughly before and after having contact with each patient.

- All patients must wash their hands thoroughly.

- All surgical instruments must be sterilised before use.

- All hospital wards must be cleaned regularly with antibacterial cleaner.

- Doctors and surgeons must wear disposable face masks, gowns and gloves.

- All spillages of blood, bodily fluids, vomit, etc. must be cleared up immediately.

- All patients with infectious diseases must be isolated to prevent the disease spreading.

How Science Works

You need to be able to explain how the treatment of disease has changed as a result of increased understanding of the action of antibiotics and immunity...

... and evaluate the consequences of mutations of bacteria and viruses in relation to epidemics and pandemics, e.g. bird influenza.

Example

2006

Genetically Modified Vaccine may be Answer to Bird Flu

Currently, scientists are trying to prevent an epidemic (large scale infection) of bird influenza becoming pandemic (a worldwide infection) by developing new vaccinations. There are fears that the bird flu virus could mutate into a strain that can be transmitted between humans. US scientists are currently working on a genetically engineered vaccine that has been found to protect mice from the strains of bird flu that recently killed people in Asia and Europe. The vaccine, which was created from a genetically modified common cold virus, has been shown to stimulate the white blood cells to produce specific antibodies that may fight a number of strains of the bird flu virus.

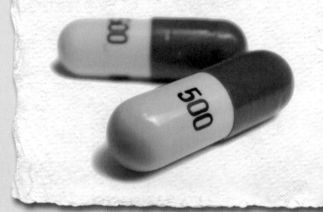

Treatment of disease changed drastically with the understanding of antibiotics and immunity.

Antibiotics are medicines which kill bacteria, so many bacterial infections and diseases, e.g. bronchitis and malaria, can be treated quite simply with a course of antibiotics. Patients suffering from illnesses which could once have led to death, now have a good chance of recovery through use of antibiotics.

People can acquire immunity to many diseases through vaccination, which means many, once common diseases, can now be prevented. Today, most young children are given vaccinations to immunise them from diphtheria, whooping cough, polio, meningitis and tuberculosis. Most are also given the MMR vaccine to immunise them against measles, mumps and rubella.

Vaccines are also available for tourists visiting places such as some parts of Africa and Asia, to protect them from diseases such as rabies and yellow fever. These diseases and illnesses could be fatal without modern medicine.

So what are the new problems facing today's scientists researching the prevention and cure of disease?

- Some strains of bacteria have developed resistance to antibiotics as a result of natural selection, e.g. MRSA (Methicillin-resistant Staphylococcus aureus). This results in the need for stronger antibiotics to be developed.
- Some bacteria and viruses have mutated (changed their form) so that existing drugs are no longer effective. The influenza (flu) virus can change rapidly, so a vaccine which combated its effect on the body one year will no longer be effective the following year. This means scientists are always having to develop new vaccines and find new ways to protect against these new strains.

You need to be able to evaluate the advantages and disadvantages of being vaccinated against a particular disease.

Example

issues

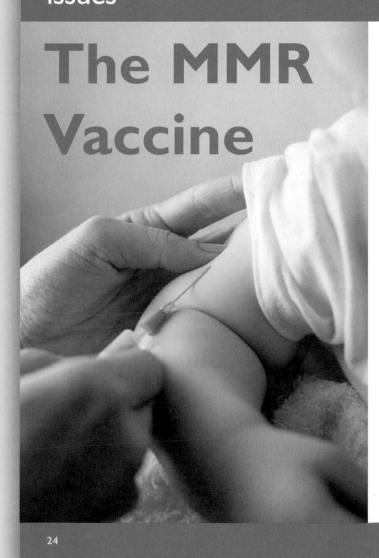

The MMR Vaccine

In Japan, scientists undertook a study of over 30,000 children born between 1988 and 1996 in the city of Yokohama. The triple vaccine (MMR) was made compulsory in 1989 and withdrawn in 1993, so they were able to compare how it affected those children who had been given the MMR vaccine with those who had not.

They counted the number of children who were diagnosed as being autistic by the age of 7. Their findings showed that for the children born in 1990, the number of cases was 86 per 10 000 children. For the children born in 1991, the figure was 56 per 10 000 children.

When the MMR was withdrawn in 1993, the children were given separate vaccinations. For children born in 1994, the number of cases of autism was 161 per 10 000.

As the number of cases of autism rose after the MMR vaccine was withdrawn, the scientists concluded that the MMR vaccine was not responsible.■

24

Benefits of Vaccination against MMR	Problems with Vaccination against MMR
• The MMR vaccine immunises children against three potentially fatal diseases – mumps, measles and rubella. • The widespread use of the MMR vaccine prevents an epidemic of mumps, measles or rubella. • The MMR vaccination means children only need one jab, rather than three separate ones.	• Smaller studies have suggested a link between the MMR vaccine and autism in children. • Some larger studies do not rule out that MMR may trigger autism in a small number of children.

Unit 1

11.5

What determines where particular species live and how many of them there are?

Species adapt to live in their surrounding environments. How many live in a certain area depends on certain factors. To understand this, you need to know…

- what organisms need in order to survive
- how plants and animals compete
- how plants and animals adapt.

To survive, organisms need a supply of materials from their surroundings and from the other organisms there.

Competition

Organisms compete with each other for space / light, food and water.

Factor	Plants	Animals
Space / light	Need room to spread leaves and obtain light for photosynthesis.	Need space to breed and compete for a mate. Also territory to hunt in.
Food	Absorb nutrients from the soil.	Herbivores compete for vegetation and carnivores compete for their prey.
Water	Absorb water by their roots.	Need water in order to survive.

Animal populations are also affected by predators, disease and migration.

Plant populations are also affected by grazing by herbivores, and disease.

A **population** is the total number of individuals of the same species which live in a certain area, e.g. the number of field mice in a meadow. A **community** is all the organisms in a particular area, i.e. many populations of plants and animals.

When organisms compete in an area or habitat, those which are better adapted to the environment are more successful and usually exist in larger numbers, often resulting in the complete exclusion of other competing organisms.

Adaptations

Adaptations are special features or behaviour which make an organism particularly well-suited to its environment. They are part of the evolutionary process which 'shapes life' so that a habitat is populated by organisms which excel there.

Adaptations increase an organism's chance of survival; they are biological solutions to an environmental challenge. For example, some plants (e.g. roses and cacti) have thorns to prevent animals from eating them. Other organisms (e.g. blue dart frogs) have developed poisons and warning colours to deter predators.

Below is an example of another organism that is well suited to its environment.

Life in a Very Cold Climate – the Polar Bear
- Rounded shape means a small surface area / volume ratio to reduce heat loss.
- Large amount of insulating fat beneath the skin, which also acts as a food store.
- Thick greasy fur to add to insulation against the cold, and to repel water.
- White coat so it is camouflaged.
- Large feet to spread its weight on the ice.
- Powerful swimmer so that it can catch its food.

You need to be able to suggest how organisms are adapted to the conditions in which they live.

Example

A group of students from a school in North Yorkshire set off to the coast to investigate the effect of waves on organisms on the seashore.

They decided to base their study on the width and height of limpets in two rocky bays – Runswick Bay and Robin Hood's Bay – to see how they were adapted to their environment.

The rocks at Runswick Bay were large boulders. They were exposed to the force of the waves on one side but sheltered on the other side.

At Robin Hood's Bay, the sandstone outcrops were shelved as they stretched gradually out to sea.

RUNSWICK BAY

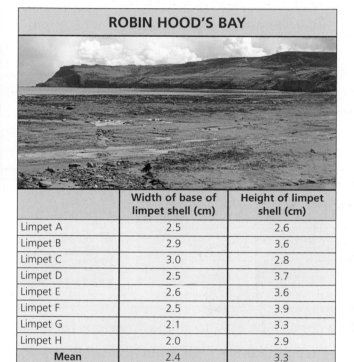

	Width of base of limpet shell (cm)	Height of limpet shell (cm)
Limpet 1	1.4	1.4
Limpet 2	2.4	1.5
Limpet 3	2.5	1.6
Limpet 4	2.4	2.5
Limpet 5	2.6	2.6
Limpet 6	3.5	2.1
Limpet 7	1.3	2.1
Limpet 8	2.3	1.9
Mean	2.3	2.0

ROBIN HOOD'S BAY

	Width of base of limpet shell (cm)	Height of limpet shell (cm)
Limpet A	2.5	2.6
Limpet B	2.9	3.6
Limpet C	3.0	2.8
Limpet D	2.5	3.7
Limpet E	2.6	3.6
Limpet F	2.5	3.9
Limpet G	2.1	3.3
Limpet H	2.0	2.9
Mean	2.4	3.3

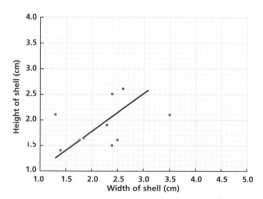

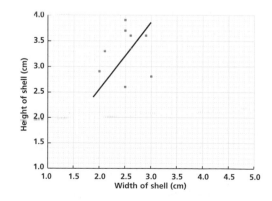

The data showed that at Runswick Bay, where the boulders are exposed to the waves, the limpet shells had a broad base compared to their height. This allowed them to resist the action of the waves.

The data from Robin Hood's Bay showed that the limpet shell bases were quite narrow compared to the height. The wave action here is not as forceful as at Runswick Bay so the limpet does not require such a broad base to hold it on the rocks.

Unit 1

You need to be able to suggest the factors for which organisms are competing in a given habitat.

Different species (e.g. plants, animals) can compete for certain factors. Different organisms of the same species have to compete against each other too.

Example 1: Red and Grey Squirrels need…	Example 2: Seaweed needs…
• **space:** red squirrels' natural habitat is woodland, but much has been destroyed, reducing the space available to them. Grey squirrels also live in woodland but they have been able to adapt to other habitats such as parks and gardens. • **food:** grey squirrels and red squirrels eat the same foods. However, grey squirrels can, if necessary, live on acorns, of which there are plenty. Red squirrels cannot and so this extra food gives grey squirrels a better chance of survival. • **shelter from prey:** both red and grey squirrels spend a lot of their time high up in the trees to escape attack from foxes and birds of prey.	• **light for photosynthesis:** many species of seaweed are found in shallow waters because they need light for photosynthesis and sunlight can only penetrate water up to a certain depth. • **water and nutrients:** unlike most other plants, seaweed absorbs water and nutrients through its surface, not through its roots. This is why it needs to spend most of its time submerged. • **a 'holdfast' (somewhere to anchor):** because it doesn't have roots, seaweed needs to anchor itself on to a rock, or something similar, to avoid being swept out to sea.

    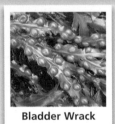 

Red Squirrel   **Grey Squirrel**   **Spiral Wrack**   **Knotted Wrack**   **Bladder Wrack**   **Oar Weed**

You need to be able to suggest reasons for the distribution of animals or plants in a particular habitat.

Some species adapt so that they can survive in conditions where their competitors can't. This affects the distribution of organisms in a particular habitat.

Example

Species of Seaweed	Habitat	Reasons for distribution
Spiral wrack	Shallow water with long periods of exposure on the upper shore.	Contains more water when hydrated and loses it slowly. It does not have air bladders.
Knotted wrack	Mid-shore, spends less time exposed to the air.	Has single bladders and varying lengths of thallus (stalks). It contains less water.
Bladder wrack	The surface of the sea.	Is often covered so has less mucus and less water initially contained in it. It has double bladders to allow it to float.
Oar weed	The surface of the sea.	Cannot withstand drying out. It has very long thallus (stalks) and large flat fronds (leaves) like hands which are lifted by the tide.

11.6

Why are individuals of the same species different from each other? What new methods do we have for producing plants and animals with the characteristics we prefer?

Differences are due to inheritance and environment. Modern science can alter, add or remove genes to produce favourable characteristics. To understand this, you need to know about...

- genes in chromosomes
- sexual and non-sexual reproduction
- how plant cuttings produce identical plants
- modern cloning techniques
- genetic engineering and why it is used.

The Genetic Information

The nucleus of a cell contains **chromosomes**, which are made up of a substance called **DNA**. A section of a chromosome is called a **gene**. Genes carry information that controls the characteristics of an organism. Different genes control the development of different characteristics. During reproduction, genes are passed from parent to offspring (i.e. they are inherited).

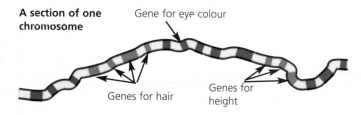

A section of one chromosome
Gene for eye colour
Genes for hair
Genes for height

Chromosomes come in pairs, but different species have different numbers of pairs, e.g. humans have 23 pairs. This example has just two pairs:

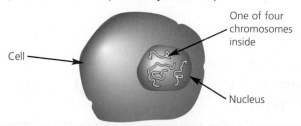

Cell
One of four chromosomes inside
Nucleus

Causes of Variation

Differences between individuals of the same species are called **variation**. Variation may be due to genetic causes (i.e. the different genes that have been inherited) or environmental causes (i.e. the conditions in which the organism has developed). Usually variation is due to a combination of genetic and environmental causes.

For example, identical twins will look exactly alike at birth, but their lifestyle can alter how they look. So if one twin has a diet high in fat and does no exercise, he will become fatter than his brother. This is an example of environmental factors. Genetics are responsible for things such as the colour of dogs'

coats.

Effect of Reproduction on Variation

During **sexual reproduction** a sperm from a male fuses with an egg from a female. When this happens, the genes carried by the egg and the sperm are mixed together to produce a new individual.

This process is completely random, which produces lots of variation, even amongst offspring from the same parents.

Asexual reproduction means no variation at all, unless it is due to environmental causes.

Only one parent is needed and individuals who are genetically identical to the parent (clones) are produced. Bacteria reproduces asexually.

Unit 1

Reproducing Plants Artificially

Plants can reproduce **asexually** (i.e. without a partner) and many do so naturally. All the offspring produced asexually are **clones** (i.e. they are genetically identical).

Taking Cuttings

When a gardener has a plant with all the desired characteristics he may choose to produce lots of them by taking stem, leaf or root cuttings. These should be grown in a damp atmosphere until roots develop.

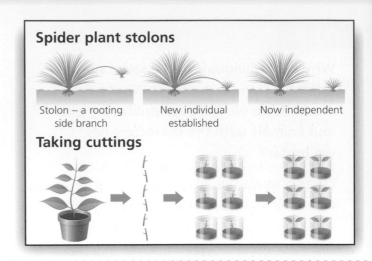

Spider plant stolons

Stolon – a rooting side branch | New individual established | Now independent

Taking cuttings

Cloning

Clones are genetically identical individuals. For example, if you have a plant which has the ideal characteristics you can clone it to produce more plants with the same desired characteristics. This is exactly what is happening in modern agriculture.

Tissue Culture

1. Parent plant with the characteristics that you want.
2. A few cells are scraped off into several beakers containing nutrients and hormones.
3. A week or two later there are lots and lots of genetically identical plantlets growing. The same can be done to these.
4. This whole process must be aseptic (carried out in the absence of harmful bacteria) otherwise the new plants will rot.

Note, the offspring are genetically identical to each other and to the parent plant.

Embryo Transplants

Instead of waiting for normal breeding cycles farmers can obtain many more offspring by using their best animals to produce embryos which can be inserted into 'mother' animals.

1. Parents with desired characteristics are mated.
2. Embryo is removed before the cells become specialised...
3. ...then split apart into several clumps.
4. These embryos are then implanted into the uteruses of sheep who will eventually give birth to clones.

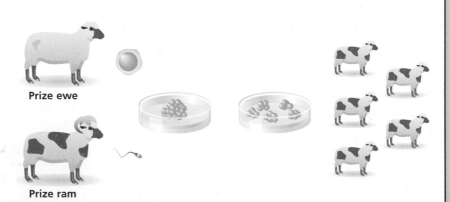

Prize ewe

Prize ram

Note, the offspring are genetically identical to each other, but not to the parents.

Adult Cell Cloning / Fusion Cell Cloning

1. DNA from a donor animal is inserted into an empty egg cell (nucleus removed)...
2. ...which then develops into an embryo.
3. The embryo is then implanted into the uterus of another sheep. Dolly the sheep was produced this way.

Reasons for Genetic Modification

Reasons for altering an organism's genetic make-up are...

- to improve the crop yield, e.g. to produce larger tomatoes, potatoes, wheat seed-heads, more oil from oilseed rape, etc.
- to improve resistance to pests or herbicides, e.g. soya plants have been modified so that they are resistant to herbicides, allowing farmers to eliminate weeds without killing the crop
- to extend the shelf-life of fast-ripening crops such as tomatoes
- to harness the cell chemistry of an organism so that it produces a substance that you require, e.g. production of human insulin.

All these processes involve transferring genetic material from one organism to another. In animals and plants, genes are often transferred at an early stage of their development so that the organism develops with desired characteristics. These characteristics can then be passed onto the offspring if the organism reproduces asexually or is cloned.

Genetic Engineering

Insulin is the hormone that is produced by the pancreas and helps to control the level of glucose in the blood. Diabetics can't produce enough insulin and often need to inject it. **Genes** such as the one for **human insulin** can be produced by genetic engineering, where genes from the chromosomes of humans and other organisms are cut out using enzymes.

1. Scientists use enzymes to cut a chromosome at specific places so they can remove the precise piece of DNA they want. In this case, the gene for insulin production is cut out.

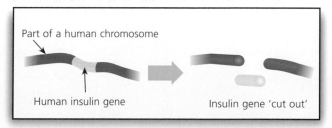

2. Another enzyme is then used to cut open a ring of bacterial DNA (a plasmid). Other enzymes are then used to insert the piece of human DNA into the plasmid.

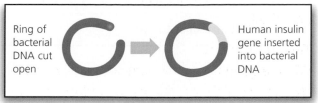

3. The plasmid is reinserted into a bacterium which starts to divide rapidly. As it divides it replicates the plasmid and soon there are millions of them, each with instructions to make insulin.

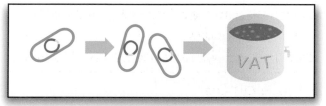

When the above process has been completed, the bacteria are cultured on a large scale and commercial quantities of insulin are then produced.

The Great Genetics Debate

Scientists have made great advances in their understanding of genes.

- They have identified genes that control certain characteristics.
- They can determine whether a person's genes may increase the risk of them contracting a particular illness, e.g. breast cancer.
- They may soon be able to 'remove' faulty genes and therefore reduce genetic diseases.

However, some people are concerned that...

- unborn children will be genetically screened and then aborted if their genetic make-up is faulty
- parents may want to choose the genetic make-up of their child
- some insurance companies may genetically screen applicants and refuse to insure people who have an increased genetic risk of an illness or disease. This may prevent these people being able to drive cars or buy homes due to lack of insurance.

How Science Works

You need to be able to make informed judgements about the economic, social and ethical issues concerning cloning and genetic engineering, including GM crops.

Argument for GM Crops	Argument against GM Crops
• They are more cost-effective (manufacturers claim higher yields). • They reduce pesticide use (according to a US study). • They can benefit human health (they can be enriched with nutrients). • They are safe for human consumption (according to the British Medical Association). • They could help the developing world by increasing yields. • They preserve natural habitats as less land is needed for agriculture.	• They increase pesticide use as farmers spray freely (according to a US study on maize). • Cross-contamination of non-GM crops could destroy the GM-free trade. • They mainly benefit big GM companies. • Unknown long-term health risks of antibiotic resistance (there is no actual evidence for this). • Increasing yields will not help the developing world (the problem is distribution of food, not lack of it). • Could affect wildlife since there are no weeds as a food source for animals.

Argument for Cloning Plants, Animals & Humans	Argument against Cloning Plants, Animals & Humans
• Traditional breeding methods are slow. • Can quickly predict characteristics of offspring. • Allows quick response to livestock / crop shortages. • Produces genetically superior stock. • Elimination of diseases such as diabetes. • Organ donation: a clone has matching tissue to the parent, so it would be able to donate an organ without the risk of the receiver rejecting the organ.	• Loss of livelihood by traditional farmers. • Risk of expression of 'unwanted' genes which adversely affect stock. • May cause backlash against cloned stock leading to market crashes. • Cloning companies may have monopoly on patent for clones. • Cloning is unnatural. • The fear of creating the 'perfect race'. • Human clones will not have 'parents'. • Cloning goes against the principles of some religions. • Abnormalities may occur in a clone. • Cloning does not allow natural evolution.

11.7

Why have some species of plants and animals died out? How do new species of plants and animals develop?

Some species die out due to environmental changes. Other species survive better due to alterations in their genes which can sometimes result in a new species. To understand this, you need to know about…

- the theory of evolution
- similarities and differences between species
- why certain species have become extinct
- evolution and natural selection
- gene mutation.

The Theory of Evolution

The theory of **evolution** states that all living things which exist today, and many more that are now extinct, evolved from simple life forms which first developed 3 000 000 000 (3 billion) years ago.

Studying the similarities and differences between species can help us to understand evolution.

The Reasons for Extinction of Species

- Increased **competition** – Australian limpets were out-competed by British limpets and are now extinct.
- Changes in the **environment** – the mammoth was once well-adapted to live in the cold environment, but became poorly adapted for the current global climate.
- New **predators** – the dodo was hunted by humans, and animals introduced by humans.
- New **diseases**.

The Fossil Record

Fossils are the remains of plants or animals from many years ago which are found in rock. They provide evidence of how organisms have changed over time.

If we look at exposed rock strata, it is possible to follow the gradual changes which have taken place in an organism over time. Even though the fossil record is incomplete, these gradual changes confirm that species have changed over long periods of time providing strong evidence for evolution.

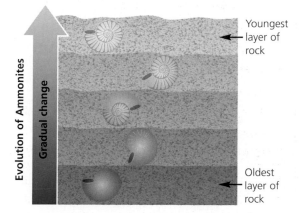

Evolution by Natural Selection

Evolution is the change in a population over a large number of generations that may result in the formation of a new species, the members of which are better adapted to their environment. There are five key points to remember.

1. Individuals within a population show variation (i.e. differences due to their genes).
2. There is competition between individuals for food and mates etc., also predation and disease keep population sizes constant in spite of the production of many offspring, i.e. there is a struggle for survival, and many individuals die.
3. Individuals better adapted to the environment are more likely to survive, breed successfully and produce offspring. This is 'survival of the fittest'.
4. These survivors will pass on their genes to their offspring resulting in an improved organism being evolved through **natural selection**.
5. Where new forms of a gene result from **mutation** there may be a more rapid change in a species.

How Science Works

You need to be able to...

- identify the differences between Darwin's theory of evolution and conflicting theories
- suggest reasons why Darwin's theory of natural selection was only gradually accepted
- suggest reasons for the different theories and why scientists cannot be certain about how life began on Earth.

Darwin's Theory

Charles Darwin sailed around the world in the 1830s collecting evidence for his 'theory of natural selection'. This states that...

- there is much variation in a species
- more offspring are produced than the environment can support (for example, food and breeding areas)
- only the ones best suited or adapted will survive and breed. This is known as natural selection or 'survival of the fittest'
- those that survive pass their genes onto their offspring. Eventually the species which is less suited will become extinct.

Darwin's theory was only gradually accepted because...

- religion had an important place in society
- it is difficult to prove
- many scientists didn't (and some still don't) accept the theory
- attempts to demonstrate evolution through tests have failed.

The Conflicting Theories

1. **Creationist**
   This view states that each living thing was created separately and did not evolve. They view the gaps in the fossil record as evidence against evolution.

2. **Lamarck's Theory**
   Jean Baptiste Lamarck thought that living things changed throughout their lives and these changes were passed on to their offspring. So he had the idea that giraffes grew long necks to reach the highest leaves on a tree. He would have also believed that if a person learnt to play the piano, that person's children would be born able to play the piano. His was 'the theory of inheritance of acquired characteristics'.

3. **Intelligent Design (ID)**
   This view states that certain structures within cells, such as DNA and mitochondria, are far too complicated to have evolved over time so must have been put there by some other higher being or creator.

Reasons for the different theories may include...

- **religion:** people who are religious may not accept scientific theories because they believe in a creator
- **culture:** people's backgrounds can influence the way they think
- **evidence:** certain theories may have more evidence to support them than others
- **knowledge:** people believe what they know
- **status of theorists / scientists:** people may be more likely to believe the ideas of renowned scientists or prominent people.

Remember that there are many different theories and scientists cannot be absolutely certain about how life began on Earth because it is difficult to find evidence to prove any theory, and theories are based on the best evidence available at that time.

No-one experienced the beginning of life on Earth so it is impossible to ever be certain how it began. Even today, we are still finding out about things and developing our scientific knowledge.

You need to be able to interpret evidence relating to evolutionary theory.

Example

Sea Monster Fossils Discovered

Fossil evidence has been discovered of what appears to be a 'sea monster', thought to be 135 million years old.

Dakosaurus' serrated tooth

The animal's large skull was found in an area in Argentina that was once part of the Pacific Ocean.

Measuring four metres (13 feet) in length, the *Dakosaurus* had four paddle-like limbs, thought to be used for balance. It had a tail like a fish which propelled it through the sea, and a head that bore a resemblance to a carnivorous dinosaur.

The fossil displays a short, high snout and big, serrated teeth similar to a crocodile's.

Every other known marine reptile had a long snout and sharp, identical teeth, which were used to deal with their main prey – small fish.

Dakosaurus's unusual snout, sharp, jagged teeth and huge jaws suggest that it hunted for sea reptiles and other large marine creatures, using its teeth to bite into and chop up its prey.

Identify and Interpret the Evidence Collected by the Scientists

1 Question

Was Dakosaurus related to today's crocodiles?

2 Evidence

For crocodile

- Large skull like a crocodile.
- Short, high snout similar to a crocodile.
- Serrated (jagged) teeth like a crocodile.

Against crocodile

- Tail like a fish.
- Paddle-like limbs, no legs.
- Large jaws and teeth suggest it preyed on large sea reptiles and other sea creatures.

3 Interpretation

- *State your view*
  Dakosaurus was a sea monster related to today's crocodiles.
- *Explain your view*
  This seems to be true because the fossils were similar to crocodiles – they had a short, high snout and serrated teeth.
- *Conclude*
  The characteristics are not found in marine reptiles. This means that Dakosaurus was a reptile with many similarities to today's crocodiles. However, unlike crocodiles, Dakosaurus was adapted to live in the sea as a sea creature.

Unit 1

11.8

How do humans affect the environment?

Humans affect the environment in many ways, resulting in changes to natural ecosystems and populations of species. Both the local and global environment may be harmed permanently as the human population increases. To understand this, you need to know…

- how the increasing human population results in pollution, waste and loss of raw materials
- how pollution can be indicated by living organisms
- what the effects of deforestation are
- how the atmosphere is affected by increasing levels of carbon dioxide and methane
- why there is a need for sustainable development and planning.

The Population Explosion

The standard of living for most people has improved enormously over the past 50 years and developments in science and medicine mean that people are now living for longer. The human population is therefore increasing exponentially (i.e. with accelerating speed).

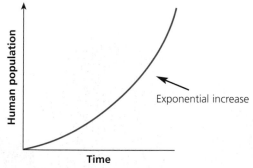

The rapid increase in the human population causes the following major problems…

- raw materials, including non-renewable energy resources, are being mined and quarried more intensively than ever before and they are being used up
- more and more domestic and industrial waste is being produced which means more landfill sites are needed

- improper handling of waste is leading to an increase in environmental pollution
- more land is taken up by farms to grow crops and keep animals
- towns and cities are expanding which means there is less land available for plants and animals.

Pollution

Human activities may pollute water, air and land:

- **water** – with sewage, fertiliser or toxic chemicals.
- **air** – with smoke and gases such as carbon dioxide, sulfur dioxide and oxides of nitrogen, which may contribute to acid rain.
- **land** – with toxic chemicals such as pesticides and herbicides, which may be washed from land into water.

Unless waste is properly handled and stored, more pollution will be caused.

Indicators of Pollution

Living organisms can be used as measures or indicators of pollution, for example…

- lichens (a blend of algae and fungus) can indicate air pollution
- the presence or absence of invertebrate animals can indicate water pollution, e.g. freshwater shrimp survive only in unpolluted water.

Deforestation

Deforestation involves the large-scale cutting down of trees for timber, and to provide land for agricultural use. This has occurred in many tropical areas with devastating consequences for the environment.

- Deforestation has increased the amount of CO_2 (carbon dioxide) released into the atmosphere due to the burning of the wood and also as a result of the decay of the wood by microorganisms.
- Deforestation has also reduced the rate at which carbon dioxide is removed from the atmosphere by photosynthesis.
- Deforestation reduces biodiversity and results in the loss of organisms that could be of future use.

Sustainable Development

It is important to improve the quality of life on Earth without compromising future generations. This is **sustainable development** and it is needed at local (towns), regional (countries) and global (worldwide) levels.

The Greenhouse Effect

The **Greenhouse Effect** describes how gases, such as methane and carbon dioxide, act like an insulating blanket by preventing a substantial amount of heat energy from 'escaping' from the Earth's surface into space. Without this effect the Earth would be a far colder and quite inhospitable place. However, the levels of these gases are slowly rising and so too is the overall temperature of the planet.

A rise in temperature by only a few degrees Celsius may lead to substantial climate changes and a rise in sea level.

- Methane and carbon dioxide gases reduce the amount of heat radiated into space.
- Deforestation reduces photosynthesis which removes CO_2.
- Burning the chopped-down wood, and industrial burning, produces CO_2.
- Increased microorganism activity on decaying material produces CO_2.
- Herds of cattle and rice fields produce methane, CH_4.

All of the above cause an increase in atmospheric carbon dioxide and methane which causes global warming.

Sustainable Development (cont.)

Sustainable development is concerned with three related issues:

- economic development
- social development
- environmental protection.

The United Nations Earth Summit in Rio de Janeiro in 1992 was arguably the first major event that resulted in a coordinated worldwide effort to produce sustained economic and social development that would benefit all the world's people,

particularly the poor, whilst balancing the need to protect the environment by reducing pollution and ensuring resources are sustainable.

Sustainable resources are resources that can be maintained in the long term at a level that allows appropriate consumption or use by people. This often requires limiting exploitation by using quotas or ensuring the resources are replenished / restocked.

The Johannesburg Summit in 2002 led to the United Nations increasing efforts to support new developments whilst sustaining the planet.

Example 1: Cod in the North Sea

The UK has one of the largest sea fishing industries in Europe. To ensure the industry can continue and fish stocks can be conserved, quotas are set to prevent over-fishing.

In 2006 the European Union Fisheries Council made changes which included…

- increasing mesh size to prevent young fish being caught before they reach breeding age
- increasing quotas of certain other types of fish other than cod.

Example 2: Pine Forests in Scandinavia

Scandinavia uses a lot of pine wood to make furniture and paper, and to provide energy. To ensure the long-term economic viability of pine-related industries, companies replenish and restock the pine forests by planting a new sapling for each mature tree they cut down.

Endangered Species

When countries or companies neglect the ideas of sustainable development, various species can become **endangered**. Below are some examples.

- The red kite (bird of prey) was exploited for its feathers.
- The numbers of osprey were reduced as its habitats were destroyed.
- The red squirrel was endangered when the larger grey squirrel was introduced.

Many endangered species are now protected. For example, the Countryside Council for Wales provides legal protection for red squirrels; they cannot be trapped, killed or kept except under special licence. The red kite and osprey both have protected sites in Wales where they can live and breed undisturbed.

Education has become a powerful 'weapon' in protecting endangered species and promoting the ideas behind sustainable development.

You need to be able to analyse and interpret scientific data concerning environmental issues.

Example

Green Week
F A C T S H E E T

How pollution affects the environment

Sulfur dioxide (SO_2) and nitrogen dioxide (NO_2) are pumped into the atmosphere by power stations and industrial plants.

Once in the atmosphere, they dissolve in water vapour to form sulfuric acid and nitric acid.

These fall to the Earth as acid rain, which alters the pH of rivers and lakes.

Human activity often upsets the natural balance of ecosystems, or changes the environment so that some species find it difficult to survive. The presence or absence of certain species acts as a pollution indicator.

Acid rain has been a problem ever since the industrial revolution began. The main gases involved are sulfur dioxide (SO_2) and nitrogen dioxide (NO_2) which are emitted from power stations and industrial plants into the air. Here they dissolve in water vapour to form acids which fall to the Earth as acid rain.

Acid rain will affect the pH of rivers and lakes. Some aquatic creatures cannot survive in acidic conditions (i.e. pH levels less than 7).

Others, however, are more tolerant and can survive at quite low pH levels (i.e. highly acidic levels).

The chart opposite shows the levels of acidity that can be tolerated by four different aquatic species: freshwater mussels, mayfly nymphs, water boatmen, and frogs.

The presence and absence of certain 'indicator species' can provide evidence of pollution both in the recent past, and at present. For example...

- the presence of water boatmen alone indicates quite high acidity (pH 3.5–5)
- the presence of frogs and the absence of freshwater mussels indicate some acidity (pH 5–6)
- the presence of frogs and mayfly nymphs, and the absence of freshwater mussels indicate some acidity (pH 5.5–6).

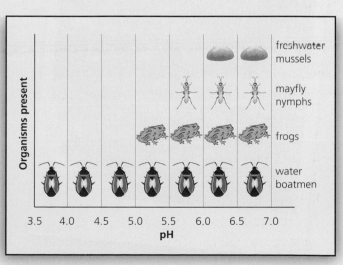

How Science Works

You need to be able to evaluate methods used to collect environmental data and consider their validity and reliability as evidence for environmental change.

Example

A power station was built upstream from a farmer's land. He was concerned about the effect that the power station was having on the local environment, so he contacted a team of scientists to carry out an investigation on the stream.

The scientists carried out a survey at the point where the shallow stream left the farmer's land. They decided to find out what the predominant species was in the stream to use as an indicator of the pH levels in the stream. They used the sweep net method to collect data.

The scientists found that the bloodworm was the predominant species. This indicates that the quality of the water was quite low. Bloodworms can tolerate poor water quality, whereas other species may be more sensitive to water quality and may, therefore, not be able to survive.

They concluded that the power station had reduced the quality of the water.

In this investigation the evidence gathered will be limited by the method used because...

- there was no control group to compare the results to – they only carried out the investigation at one point in the stream, and they did not know what the pH levels were before the power station was operating
- they carried out their investigation at the point where the stream left the farmer's land, so it did not taken into account the effect of the chemicals the farmer might use on his crops
- the sweep net method is not the most appropriate sampling method for a shallow stream
- there is no mention of how they determined what the predominant species was
- although bloodworms can survive in low quality water, they can also survive in better quality water, so it is not a very reliable measure
- it does not state how many times they repeated the experiment (if at all)
- other variables that might influence the results (e.g. season, weather) were not considered.

You need to be able to weigh up the evidence and form balanced judgements about some of the major environmental issues facing society, including the importance of sustainable development.

Example

International Whaling Commission

Home

Newsroom

News Archive

About Whales

Shop

Whale Watching

Kids' Corner

Links

Contact Us

Gallery

The plight of the oceans' whales was highlighted in January 2006 when a Northern Bottlenose whale lost its way and swam up the River Thames in London. It struggled to find its way to the sea and marine experts were asked for their opinions about what to do. Sadly it died, but not before the 'whale debate' started again.

Whales have been hunted for hundreds of years. The Minke whale has now been hunted to such an extent that it could soon be extinct. In 1986, the International Whaling Commission agreed to ban whale hunting.

Some whaling communities in countries such as Norway lost income at first but they started to offer 'whale-spotting' trips to supplement their earnings. However, some scientists argued that there was a need to kill some whales for research purposes. Others said that the whales need to be alive in order to study their communication and migration.

The ban has now been lifted and some countries claim to be hunting whales for scientific research, but make a profit from the whale carcasses.

Sustainable development is concerned with the careful management of natural resources. It allows us to continue using these resources, whilst ensuring that stocks do not run out and that adequate levels are maintained for the future. For example, where animals are being taken out of the environment for food, or other reasons, it is important to ensure that enough individuals are left to breed and maintain future populations. This means that numbers must be monitored carefully, so that appropriate action can be taken when necessary.

Your judgement may not be based on evidence alone as other social factors may be relevant. To help you make your judgement about whaling, consider the following:

- certain viewpoints may be biased, e.g. whalers rely on hunting whales for their income so they will have a biased view
- evidence can be given too much weight or dismissed too lightly according to its political significance; the evidence might be downplayed if it upsets the public, e.g. the possible extinction of some whales could be dismissed by a government whose people rely upon whaling
- the status of a scientist may influence the weight placed on the evidence, e.g. a scientist with experience or professional status who presents evidence that whales can be studied without killing them is more likely to have his case heard than a lesser-known colleague.

Example Questions

For Unit 1, you will either have to complete two objective tests (matching and multiple choice questions) or one written paper (longer, structured questions).

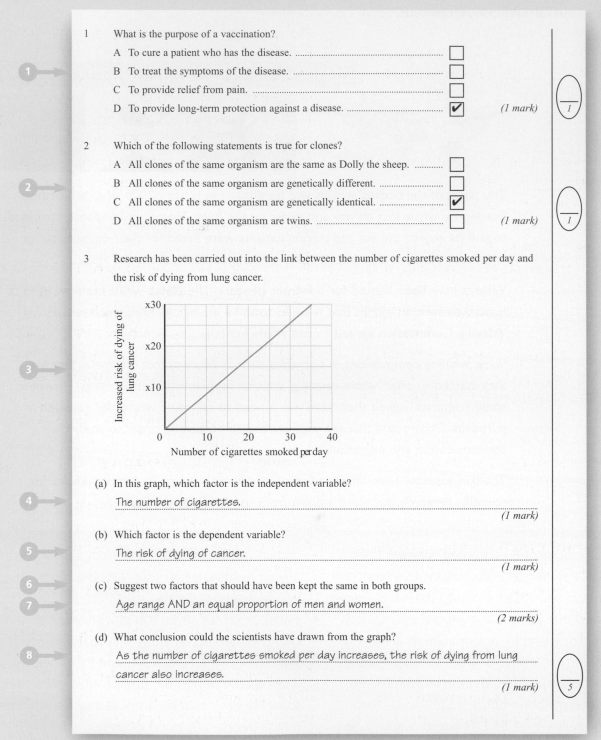

1   What is the purpose of a vaccination?

    A To cure a patient who has the disease. .... ☐

    B To treat the symptoms of the disease. .... ☐

    C To provide relief from pain. .... ☐

    D To provide long-term protection against a disease. .... ☑ *(1 mark)* ①

2   Which of the following statements is true for clones?

    A All clones of the same organism are the same as Dolly the sheep. .... ☐

    B All clones of the same organism are genetically different. .... ☐

    C All clones of the same organism are genetically identical. .... ☑

    D All clones of the same organism are twins. .... ☐ *(1 mark)* ①

3   Research has been carried out into the link between the number of cigarettes smoked per day and the risk of dying from lung cancer.

(a) In this graph, which factor is the independent variable?

    The number of cigarettes.

        (1 mark)

(b) Which factor is the dependent variable?

    The risk of dying of cancer.

        (1 mark)

(c) Suggest two factors that should have been kept the same in both groups.

    Age range AND an equal proportion of men and women.

        (2 marks)

(d) What conclusion could the scientists have drawn from the graph?

    As the number of cigarettes smoked per day increases, the risk of dying from lung cancer also increases.

        (1 mark) ⑤

---

① Questions like this often appear as part of a set of questions relating to the same subject (e.g. vaccinations).

② If you are unsure about the answer to a multiple-choice question, eliminate the options that are obviously wrong first.

③ Look at graphs carefully and make sure you understand what they show before answering any questions.

④ The independent variable is the one that is being controlled.

⑤ The dependent variable is the one that is measured each time the independent variable is changed, to see if there is a relationship between them.

⑥ Read the question carefully, this one asks for **two** factors so you would lose a mark if you only gave one.

⑦ You could name any factors that could potentially affect the dependent variable (risk of dying of lung cancer).

⑧ Describe what the data tells you – nothing else!

Key Words

Adaptation – the gradual change of a particular organism over generations to become better suited to its environment

Biodiversity – the variety among living organisms and the ecosystems in which they live

Carcinogen – a substance that produces cancer

Chromosome – a coil of DNA made up of genes, found in the nucleus of plant / animal cells

Clone – a genetically identical descendent of an organism

Deficiency disease – a disease caused by the lack of some essential element in the diet

Deforestation – the destruction of forests by cutting down trees

DNA (deoxyribonucleic acid) – nucleic acid molecules which contain genetic information and makes up chromosomes

Effector – the part of the body, e.g. a muscle or a gland, which produces a response to a sensor

Embryo – a ball of cells which will develop into a human / animal baby

Enzyme – a protein which speeds up a reaction (a biological catalyst)

Evolve – to change naturally over a period of time

Extinct – a species that has died out

Fetus – an unborn animal / human baby

Fossil – remains of animals / plants preserved in rocks

FSH (Follicle Stimulating Hormone) – stimulates ovaries to produce oestrogen

Gene – part of a chromosome, composed of DNA

Gland – an organ in an animal body used for secreting substances

Herbicide – a toxic substance used to destroy unwanted vegetation

Hormone – a regulatory substance which stimulates cells or tissues into action

Infectious – a disease that is easily spread, through water, air, etc.

Ion – minerals such as sodium and potassium, that are needed by the body

Leprosy – a contagious bacterial disease affecting the skin and nerves

LH – Luteinising Hormone, stimulates changes in the menstrual cycle

Malnourished – suffering from lack of essential food nutrients

Menstrual cycle – the monthly cycle of hormonal changes in a woman

Metabolic rate – the rate at which an animal uses energy over a given time period

Methane – a colourless, odourless, inflammable gas

MRSA – Methicillin-resistant Staphylococcus aureus (or 'superbug'), an antibiotic-resistant bacterium

Mutation – a change in the genetic material of a cell

Neurone – specialised cell which transmits electrical messages or nerve impulses

Non-renewable – resources that cannot be replaced

Obesity – the condition of being very overweight

Pathogen – an agent causing a disease

Pesticide – a substance used for destroying insects or other pests

Pituitary – a small gland at the base of the brain that produces hormones

Predator – an animal that hunts, kills and eats other animals

Receptor – a sense organ e.g. eyes, ears, nose, etc.

Recreation – for pleasure

Reflex action – an involuntary action, e.g. removing hand from hot plate

Saturated fats – animal fat, considered to be unhealthy

Sustainable – resources that can be replaced

Synapse – the gap between two nerve-cells

Toxin – a poison produced by a living organism

Vaccine – a liquid preparation used to make the body produce anti-bodies in order to provide protection against disease

Variation – differences between individuals of the same species

Unit 2

What are animals and plants built from?

All living organisms are made up of cells. The structure of each cell depends on its function. To understand this, you need to know...

- the different parts of cells, and the functions of these parts
- how chemical reactions are controlled in cells
- how cells are adapted to carry out particular functions.

Typical Plant and Animal Cells

All living organisms are made up of cells. The structures of different types of cells are related to their functions.

A palisade cell from a leaf

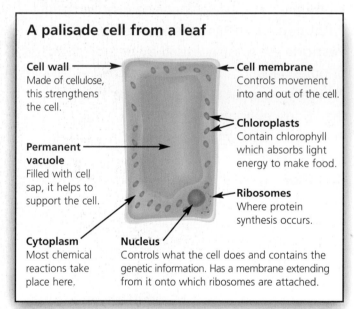

Cell wall
Made of cellulose, this strengthens the cell.

Cell membrane
Controls movement into and out of the cell.

Chloroplasts
Contain chlorophyll which absorbs light energy to make food.

Permanent vacuole
Filled with cell sap, it helps to support the cell.

Ribosomes
Where protein synthesis occurs.

Cytoplasm
Most chemical reactions take place here.

Nucleus
Controls what the cell does and contains the genetic information. Has a membrane extending from it onto which ribosomes are attached.

A cheek cell from a human

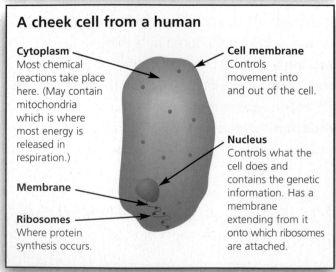

Cytoplasm
Most chemical reactions take place here. (May contain mitochondria which is where most energy is released in respiration.)

Cell membrane
Controls movement into and out of the cell.

Membrane

Nucleus
Controls what the cell does and contains the genetic information. Has a membrane extending from it onto which ribosomes are attached.

Ribosomes
Where protein synthesis occurs.

Some cells are specialised to do a particular job.

Root hair cells have a tiny hair-like structure which increases the surface area of the cell enabling it to absorb water and ions more efficiently.	
Palisade cells are column-shaped cells on the upper surface of the leaf. They are packed with chloroplasts for photosynthesis.	
Xylem cells are long, thin, hollow cells that contain no cytoplasm (they are actually dead!). They transport water through the stem and root.	
Nerve cells (neurones) have long slender axons which can carry nerve impulses over distances as long as one metre.	
The **ovum** or **egg cell** is much larger than other cells so that it can carry massive food reserves for the developing embryo.	
The **sperm cell** is the most mobile cell because of its tail. It has to travel from the vagina to the ovum.	
Red blood cells have no nucleus so that they can be packed full of haemoglobin in order to carry lots of oxygen.	Cross section
White blood cells can change their shape in order to engulf and destroy microbes which have invaded the body.	

Most cells are made up of water containing dissolved substances. These substances are usually in the process of being made into something the cell needs. This involves chemical reactions controlled by enzymes. Enzymes are found in cytoplasm and in all **mitochondria** (which produce the cell's energy).

How Science Works

You need to be able to relate the structure of different types of cells to their function in a tissue or an organ.

Example

The Lonsdale News, Saturday, June 24, 2006

+   14

environmentwatch

Deadly Fungus Threatens Frogs

The alarmingly extensive spread of a deadly fungus throughout America is causing great concern. Scientists have reported that hundreds of thousands of the much-loved golden frog, a national emblem, are dying from the effects of the fungus.

The fungus is thought to affect other amphibians as well and scientists are currently planning investigations to find out how many other species are being affected, and to what extent.

Ask Greenfingers

Dear Greenfingers,
No matter what I put into my soil – I've tried expensive nutrients and fertilisers etc. – my plants always seem to be droopy in the summer. Why does this keep happening and what can I do to prevent it this year?
Yours, Concerned of Colchester

Dear Concerned,
Sometimes, no matter how rich in nutrients your soil is, or what fertiliser you use, you can end up with droopy plants. The fact that this usually happens in the summer suggests that your plants are suffering from a lack of water. They get dehydrated just like we do, so what they need is a good watering. Ideally you should try to water your plants every day. It is best to wait until early evening to water your plants, when the heat of the sun is less powerful. I hope you have better luck with your plants this year!

Why are the frogs dying?

Frogs' skin is made of a layer of cells which have permeable cell walls. These permeable cell walls allow the passage of water and gases into the frog's blood when it is resting. This means that a frog can breathe through its skin as well as taking in water simply by jumping into a pond or sitting in a puddle.

If the skin cells are covered by an impermeable layer, such as the fungus mentioned in the article above, the cells get blocked up, which makes the exchange of liquids and gases extremely difficult and eventually the frog will die.

Why do plants droop when they do not have enough water?

Water is transported through the root and stem of a plant via long thin cells, which do not have end walls, that make up the xylem. The walls of these cells are strengthened by rings of lignin (a rigid deposit), which enable them to withstand the pressure of water. The water pressure within the xylem makes the plant turgid and strong. If there is not enough water in the cells, the cell walls become limp and flaccid and the plant droops.

Unit 2

How do dissolved substances get into and out of cells?

Dissolved substances pass through the cell membranes to get into and out of the cells. To understand this, you need to know…

- what diffusion is
- what osmosis is
- what causes osmosis.

How do plants obtain the food they need to live and grow?

Green plants make their own food using air, soil and energy from the Sun. To understand this, you need to know…

- how photosynthesis works and the factors that can slow it down
- how the glucose produced during photosynthesis is used
- what plants need to ensure healthy growth
- what happens when plants are deficient in mineral ions.

Amoeba – a single-celled organism
(not to scale)

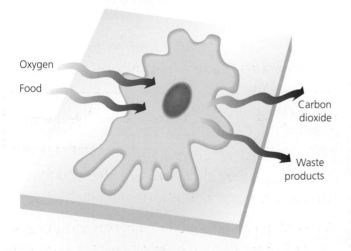

Oxygen
Food
Carbon dioxide
Waste products

Oxygen, carbon dioxide, food and waste products, along with simple sugars and ions, pass easily through cell membranes.

Dissolved substances can move into and out of cells by diffusion and osmosis.

Diffusion

Because cells are living things, they constantly have to replace substances which are used up (e.g. food and oxygen) and remove other substances which would otherwise accumulate (e.g. carbon dioxide and waste products).

Even simple, single-celled animals, like the amoeba (below left), need to do this. This can take place automatically, without the need for energy, in a process called **diffusion**.

Diffusion is the spreading of the particles of a gas, or of any substance in solution, which results in a net movement from a region where they are at a higher concentration to a region where they are at a lower concentration. For example, oxygen required for respiration passes through all membranes by diffusion.

The greater the difference in concentration, the faster the rate of diffusion. So in the example of oxygen and the amoeba, there is lots of oxygen outside the amoeba but much less inside, because it is being used up in respiration. This results in the rapid diffusion of oxygen into the amoeba through the cell membrane.

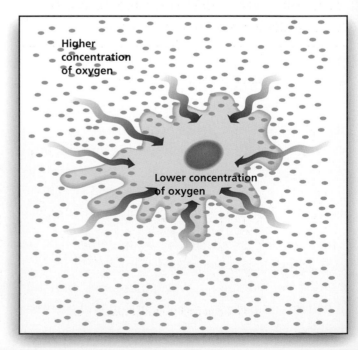

Higher concentration of oxygen

Lower concentration of oxygen

Osmosis

Osmosis is the movement of water from a dilute solution (e.g. with a high water to solute ratio) to a more concentrated solution (e.g. with a low water to solute ratio) through a partially permeable membrane. The membrane allows the water molecules through but not the solute molecules because they are too large.

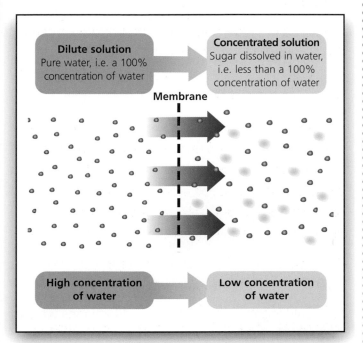

The effect of osmosis is to gradually dilute the solution. This is what happens at root hair cells, where water moves from the soil into the cell by osmosis, along a concentration gradient.

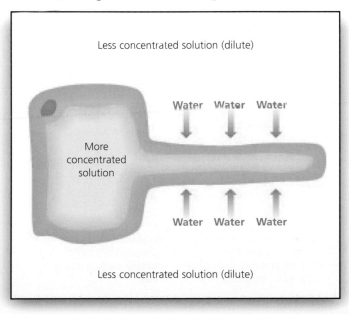

Making Food using Energy from the Sun

Green plants don't absorb food from the soil. They make their own using sunlight. This is called **photosynthesis**. It occurs in the cells of green plants, which are exposed to light.

Four things are needed...
- light from the Sun
- carbon dioxide diffused from the air
- water from the soil
- chlorophyll in the leaves.

Two things are produced...
- glucose – for biomass and energy
- oxygen – released into the atmosphere as a by-product.

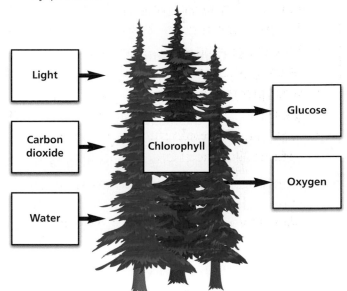

The word equation for photosynthesis is...

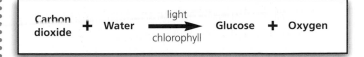

Light energy is absorbed by green chlorophyll (in chloroplasts in some plant cells).

Some of the glucose produced in photosynthesis is used immediately by the plant to provide energy via respiration. However, much of the glucose is converted into insoluble starch which is stored in the stem, leaves or roots.

Factors Affecting Photosynthesis

Temperature, carbon dioxide concentration and light intensity interact to limit the rate of photosynthesis. Any one of them, at a particular time, may be the limiting factor.

Temperature

1. As the temperature rises so does the rate of photosynthesis. This means temperature is limiting the rate of photosynthesis.
2. As the temperature approaches 45°C, the enzymes controlling photosynthesis start to be destroyed and the rate of photosynthesis drops to zero.

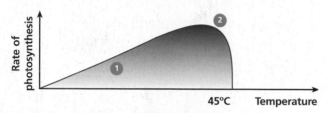

Carbon Dioxide Concentration

1. As the rate of carbon dioxide concentration rises so does the rate of photosynthesis. So carbon dioxide is limiting the rate of photosynthesis.
2. Rise in carbon dioxide levels now has no effect. Carbon dioxide is no longer the limiting factor. Light or temperature must now be the limiting factor.

Light Intensity

1. As the light intensity increases so does the rate of photosynthesis. This means light intensity is limiting the rate of photosynthesis.
2. Rise in light intensity now has no effect. Light intensity is no longer the limiting factor. Carbon dioxide or temperature must now be the limiting factor.

Photosynthesis, and therefore growth, can be controlled when these factors are controlled. For example, a greenhouse can be modified to…

- increase temperature (up to about 40°C)
- increase CO_2 levels
- increase light intensity.

This will result is plants growing more quickly and becoming bigger and stronger.

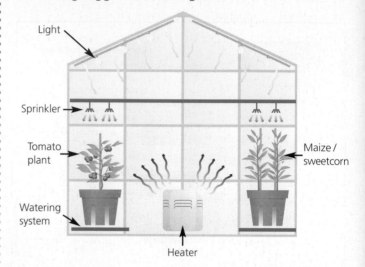

Plant Mineral Requirements

For healthy growth, plants need mineral ions which they absorb from soil through their roots, including nitrates and magnesium.

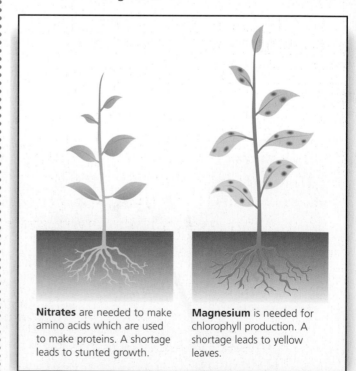

Nitrates are needed to make amino acids which are used to make proteins. A shortage leads to stunted growth.

Magnesium is needed for chlorophyll production. A shortage leads to yellow leaves.

You need to be able to interpret data showing how factors affect the rate of photosynthesis, and evaluate the benefits of artificially manipulating the environment in which plants are grown.

Example

Charlie decided to investigate the benefits of growing lettuces in artificially controlled conditions.

He used a fossil burner to investigate what effects varying the amount of carbon dioxide in the air would have on his lettuces and used weight as an indicator of growth, and therefore the rate of photosynthesis.

He divided 50 lettuces into five groups that weighed the same overall. He then adjusted the level of carbon dioxide in each group's air (the independent variable), but made sure that the temperature and the amount of light and water each group received was the same (controlled variables). After eight weeks, he weighed each group again (dependent variable). The graph above shows his results.

Conclusion

The graph shows that the more carbon dioxide in the air, the greater the mass of the lettuces. Carbon dioxide aids photosynthesis, which is why the plants that had more carbon dioxide in the air grew more than those with a lower percentage of carbon dioxide. However, it is also important to notice that the graph levels off between 4% and 5%, which would suggest that this is the optimum percentage of carbon dioxide needed for photosynthesis. Increasing the levels of carbon dioxide present in the air beyond this would not increase the rate of photosynthesis, and therefore did not affect the weight of the lettuce.

Advantage of Controlling Carbon Dioxide	Disadvantages of Controlling Carbon Dioxide
• Larger average mass of lettuces produced.	• Cost of carbon dioxide machine. • Need to monitor the levels of carbon dioxide. • Other factors could be involved that have not been taken into consideration.

Unit 2

12.4

What happens to energy and biomass at each stage in a food chain?

The relative amount of energy and biomass decreases at each stage in a food chain. To understand this, you need to know…

- what pyramids of biomass show
- how the Sun provides energy for many living organisms
- the ways in which the amounts of material and energy in the biomass of organisms is reduced at each stage in the food chain
- how the efficiency of food production in the food chain can be improved.

Food Chains

Radiation from the Sun is the source of energy for all communities of living organisms.

In green plants, photosynthesis captures a small fraction of the solar energy which reaches them. This energy is stored in the substances which make up the cells of the plant and can be passed onto organisms which eat the plant. This transfer of energy can be represented by a food chain.

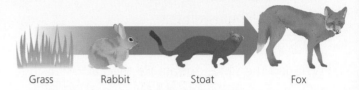

Grass　　　Rabbit　　　Stoat　　　Fox

Pyramids of Biomass

The mass of living material (**biomass**) at each stage of a food chain is less than it was at the previous stage. The biomass at each stage can be drawn to scale and shown as a pyramid of biomass.

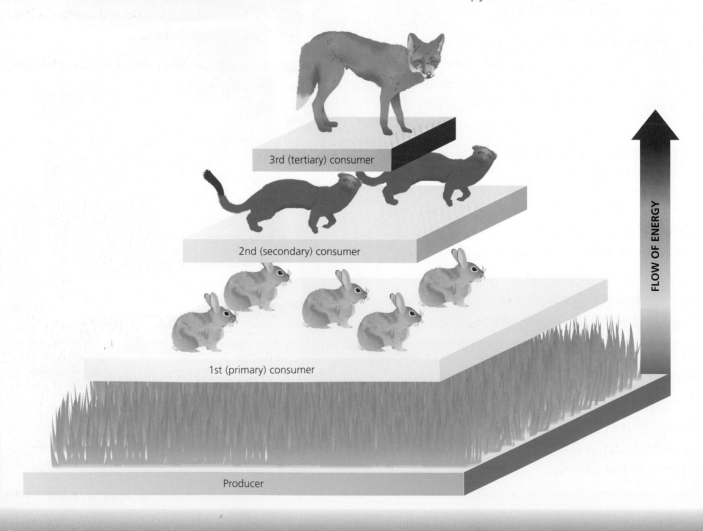

3rd (tertiary) consumer

2nd (secondary) consumer

1st (primary) consumer

Producer

FLOW OF ENERGY

Transfer of Energy and Biomass

Biomass and energy are lost at every stage of a food chain because materials and energy are lost in an organism's faeces (waste).

Energy released through respiration is used up in movement and lost as heat energy.

This is particularly true in warm-blooded animals (birds and mammals), whose bodies must be kept at a constant temperature, higher than their surroundings.

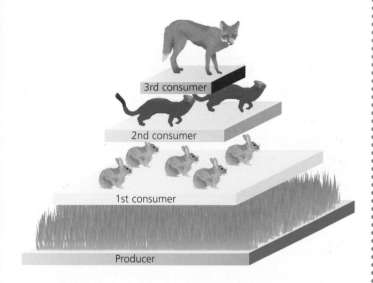

In the pyramid above, the fox gets the last tiny bit of energy and biomass that is left.

Only a fraction of the Sun's energy is captured by the producers. Much of the biomass remains in the root system and so does not get eaten.

Rabbits run, mate, excrete, generate heat and pass on only a tenth of the energy they get from lettuce. A lot of biomass is lost in droppings (faeces).

Stoats run, mate, excrete, generate heat and pass on only a tenth of the energy they get from the rabbits. A lot of biomass is lost as faeces.

Foxes, as 3rd consumers, only receive a small amount of the energy captured by the grass at the bottom of the pyramid.

Improving the Efficiency of Food Production

Since the loss of energy and biomass is due mainly to heat loss, waste and movement, it follows that we can improve the efficiency of food production by reducing the number of stages in a food chain or by limiting an animal's movement, and controlling its temperature.

…is not as efficient as…

More of the food eaten by the animal is then converted into biomass because less energy is lost through heat and movement.

However, many people feel that this way of rearing animals is unacceptable.

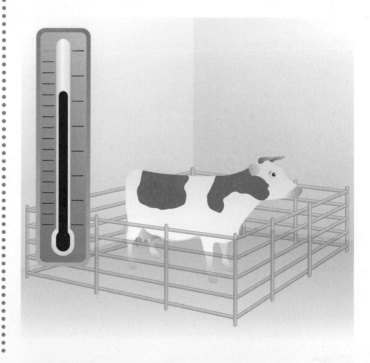

How Science Works

You need to be able to interpret pyramids of biomass and construct them from appropriate information.

Example

Species	Producer / Feeds On
Phytoplankton	Producer
Elodea (water plant)	Producer
Algae	Producer
Water beetle	Feeds on water plants
Carp	Feeds on insect larvae and snails
Pond snail	Feeds on water plants and small insects
Mosquito larvae	Feeds on insect algae and protozoa
Leech	Feeds on pond snails
Stickleback	Feeds on insect larvae and leeches

Carp (tertiary consumer)

Pond snail (secondary consumer)

Water beetle (primary consumer)

Phytoplankton (producer)

By looking at the pyramid of biomass above, we can see that it represents a pond community. It shows the relative biomass of four species that live in the pond and it also shows us what eats what in the food chain.

We can see that the producer is phytoplankton (a plant). This provides food and energy for the water beetles, which are eaten by pond snails, which in turn are consumed by carp. We can also see that the biomass of living material reduces as you go further up the pyramid (more levels of consumers). This is because biomass is lost as waste products, e.g. faeces and carbon dioxide.

The table opposite provides details of other organisms in the pond.

To construct a pyramid of biomass from this information there are a few basic rules to follow:
- it will be pyramid shaped!
- biomass is the mass of living material
- the producers are placed at the bottom
- the pyramid represents the total biomass at each stage, not the number of organisms.

Stickleback (tertiary consumer)

Leech (secondary consumer)

Pond snail (primary consumer)

Elodea (producer)

You need to be able to evaluate the positive and negative effects of managing food production and distribution, and recognise that practical solutions to human needs may require compromise between competing priorities.

Example

Farming Gazette   *June 2006*

Union Votes to Ban Battery Cages

In 1999, the European Commission proposed that the floor size of battery hen cages should be increased to a minimum of 800cm^2 (from 450cm^2). However, members of the European Parliament went even further and voted by a majority of two to one to ban battery hen cages altogether. They agreed that the ban should be enforced throughout Europe, as of 2009.

Some 93% of all eggs in the EU are produced by battery hens, so this new legislation will create a great upheaval in the European egg market.

Current Conditions

There are up to 30 million chickens in the UK; 85% live in 'battery farms' where they are kept in sheds that contain over 20,000 birds. By restricting the movement of the birds, the energy usually lost in movement is converted into meat and the chickens get bigger more quickly, which means they can be sold for meat at an earlier age. The chickens are also exposed to artificial sunrise and sunset so that egg production is increased.

A European scientific report stated that chickens kept in confined spaces can have serious health problems, including hock burn (brown marks on their legs), and pressure on the heart and legs. About 100,000 chickens die each day.

Ideal Conditions

'Free range' chickens are housed in sheds with perches but they have access to the outside during the day. However, 'free range' chickens and their eggs could be up to 25% more expensive than battery hens and their eggs.

Farmers' Concerns

Poultry farmers are worried that these changes to the industry could result in them losing money. Countries outside the EU could provide cheap imports into UK supermarkets which could be sold cheaper than UK eggs and chicken.

The Solution

In Britain, scientists were asked to provide guidelines to enable Britain to comply with the European legislation. They came up with the 'Battery Cage Rule': all battery cages are to be converted to 'enriched cages' by 2007 (enriched cages provide at least 750cm^2 floor space per bird) and all battery hen farms are to be replaced by free range and barn systems by 2012. The time delay, and the rule on 'enriched cages', was agreed in order to give farmers time to adapt their farms to comply with the new legislation.

The suggestion is also that imports of eggs and chickens should come only from countries with the same standards of housing hens as the EU. This would prevent UK poultry farmers losing out to cheap imports from outside the EU.

This is an example of a practical solution to human needs where there is a compromise between competing priorities.

Advantages of Managing Food Production
• Cheap chicken and eggs available. • Poultry farmers can own many chickens and therefore earn a good living. • Do not need lots of space. • Limiting movement increases size quicker.

Disadvantages of Managing Food Production
• Chickens are kept in poor conditions. • Many chickens die before they can be sold. • Many chickens suffer health problems.

Unit 2

12.5

What happens to the waste material produced by plants and animals?

Microbes help to decompose waste material from animals and plants. It is then used by plants as a source of nutrients. To understand this, you need to know…
- how the environment recycles waste material
- how microorganisms break down materials
- how the carbon cycle works, and its effects.

Recycling the Materials of Life

Living things remove materials from the environment for growth and other processes, but when these organisms die or excrete waste, these materials are returned to the environment.

The key to all this is the **microorganisms** which break down the waste and the dead bodies. This decay process releases substances used by plants for growth.

Microorganisms digest materials faster in warm, moist conditions where there is plenty of oxygen.

Humans also deliberately use microorganisms in…
- sewage works to break down human waste
- compost heaps to break down plant material waste.

The Carbon Cycle

In a stable community, the processes which remove materials are balanced by processes which return materials. This constant recycling of carbon is called the **Carbon Cycle**.

1. CO_2 is removed from the atmosphere by green plants to produce glucose by photosynthesis. Some is returned to the atmosphere by the plants during respiration.

2. The carbon obtained by photosynthesis is used to make carbohydrates, fats and proteins in plants. When the plants are eaten by animals this carbon becomes carbohydrates, fats and proteins in animals.

3. Animals respire releasing CO_2 into the atmosphere.

4. When plants and animals die, other animals and microorganisms feed on their bodies causing them to break down.

5. As the detritus feeders and microorganisms eat the dead plants and animals, they respire releasing CO_2 into the atmosphere.

12.6

What are enzymes and what are some of their functions?

Enzymes are produced by living cells. They are biological catalysts that have many functions inside and outside cells. To understand this, you need to know…

- what enzymes are, what they consist of, and what they do
- what processes they catalyse
- how energy released during aerobic respiration is used
- what the functions of digestive enzymes are
- how enzymes are used in the home and in industry.

Enzymes

Enzymes are biological catalysts; they increase the rate of chemical reactions in an organism.

Enzymes are protein molecules made up of long chains of **amino acids**.

Protein molecule made up of amino acids

These are folded into a 3-D shape which lets other molecules fit into the enzyme.

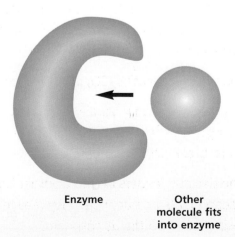

Enzyme   **Other molecule fits into enzyme**

Different enzymes work best at certain temperatures and pH levels.

High temperatures destroy most enzymes' special shape. This is why it is dangerous for a human's body temperature to go much above 37°C.

Enzyme destroyed by heat

Heat

Inside Living Cells

Enzymes in living cells **catalyse** (speed up) processes such as respiration, protein synthesis and photosynthesis.

The energy released during respiration is used to…

- build larger **molecules** using smaller ones
- enable **muscles** to contract (in animals)
- **maintain a constant temperature** (in mammals and birds) in colder surroundings
- **make proteins** in plants from amino acids (made from sugars and nitrates).

N.B. They all begin with M, so remember the four 'M's.

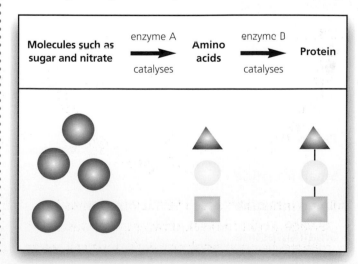

Molecules such as sugar and nitrate — enzyme A catalyses → Amino acids — enzyme B catalyses → Protein

Aerobic Respiration

Aerobic respiration releases energy through the breakdown of glucose molecules, by combining them with oxygen inside living cells. (The energy is actually contained inside the glucose molecule. Aerobic respiration mostly takes place inside mitochondria.)

The written equation for aerobic respiration is…

$$\text{Glucose} + \text{Oxygen} \xrightarrow[\text{by enzymes}]{\text{catalysed}} \text{Carbon dioxide} + \text{Water} + \text{Energy}$$

Outside Living Cells

Digestive enzymes are produced by specialised cells in glands in the lining of the gut.

The enzymes pass out of the cells into the gut where they come into contact with food molecules.

They then catalyse the breakdown of large molecules into smaller molecules.

Three enzymes – protease, lipase and amylase – are produced in four separate regions of the digestive system. They digest proteins, fats and carbohydrates to produce smaller molecules which can be absorbed.

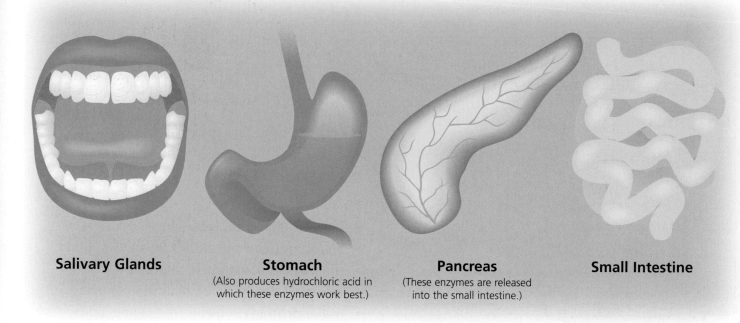

Salivary Glands

Stomach
(Also produces hydrochloric acid in which these enzymes work best.)

Pancreas
(These enzymes are released into the small intestine.)

Small Intestine

	Enzyme	What it digests	Molecules produced
Salivary glands	Amylase	Starch	Sugars
Stomach	Protease	Proteins	Amino acids
Pancreas	Amylase	Starch	Sugars
Pancreas	Protease	Proteins	Amino acids
Pancreas	Lipase	Lipids (fats and oils)	Fatty acids and glycerol
Small intestine	Amylase	Starch	Sugars
Small intestine	Protease	Proteins	Amino acids
Small intestine	Lipase	Lipids (fats and oils)	Fatty acids and glycerol

The Function of Bile

Bile is produced in the liver and then stored in the **gall bladder** before being released into the small intestine. Bile has two functions:

1 It neutralises the acid, which is added to food in the stomach, to produce alkaline conditions in which the enzymes of the small intestine work best.

2 It emulsifies fats, i.e. it breaks down large drops of fat into small droplets to increase their surface area. This enables the lipase enzymes to work much faster.

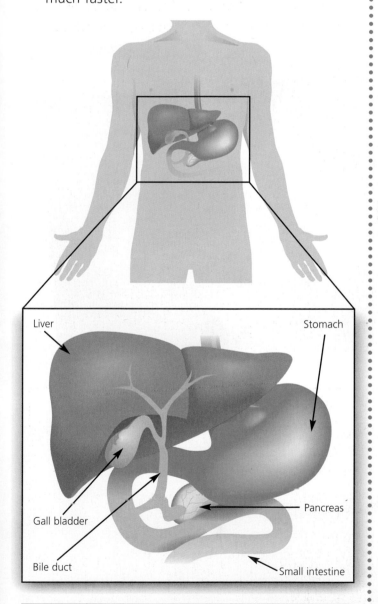

Liver
Stomach
Gall bladder
Bile duct
Pancreas
Small intestine

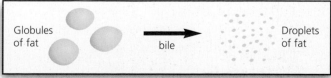

Globules of fat → bile → Droplets of fat

Use of Enzymes in the Home and Industry

Some microorganisms produce enzymes which can be used to our benefit both in the home and in industry.

In the Home
Biological detergents may contain…

* protein-digesting (protease) enzymes to break down stains such as blood and food
* fat-digesting (lipase) enzymes to break down oil and grease stains.

In Industry

* Proteases are used to pre-digest protein in baby foods.
* Carbohydrases are used to convert starch into sugar syrup.
* Isomerase converts glucose syrup into fructose syrup, which is even sweeter. It can be used in smaller quantities which makes it ideal for use in slimming foods.

How Science Works

You need to be able to evaluate the advantages and disadvantages of using enzymes in home and industry.

If you bite into an apple and leave it for a few minutes it will turn brown. The reason behind this is the unlocking of enzymes inside the cells, which catalyse a chemical reaction with the air. This results in the flesh of the fruit turning brown.

Humans have used enzymes for thousands of years in the production of food and drink, but the use of enzymes in industry is fairly new.

Enzymes are biological catalysts; they increase the rate of a reaction without harming the cells of the living organism. These enzymes are taken from microorganisms which can be grown easily and used over and over again. Many industries rely upon enzymes to speed up reactions that would normally require high temperatures or high pressure to proceed. These enzymes are often made in a fermenter, which may prove to be expensive.

However, without enzymes the reaction either would not happen or would take a long time to produce the commercial product. Some of the areas we use enzymes in are...

- biological detergents to remove protein stains (but some people develop skin allergies to them)
- medicine, for example, drug manufacture and glucose sensors for diabetics
- the leather industry to soften hides and remove hair
- making the soft centre in sweets (the sweet is injected with an enzyme to break down the sugar inside).

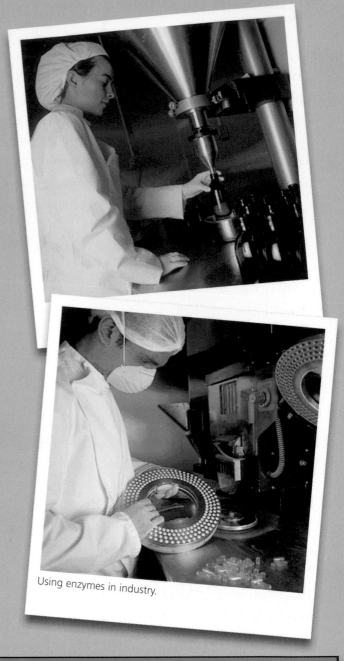

Using enzymes in industry.

Advantages	Disadvantages
<ul><li>Can be used over and over again.</li><li>Speeds up reactions.</li><li>Products are made without the need for high temperatures so they are energy saving.</li><li>Many come from microorganisms which can be easily grown.</li><li>Many have been tried and tested for hundreds of years.</li><li>Have many varied applications.</li><li>Drugs for the medical industry can be made easily on a large scale so world health is improved.</li></ul>	<ul><li>Some people are allergic to them.</li><li>Often need to be made in a large vessel called a fermenter which can be costly.</li><li>Some people don't like the idea of using microorganisms in food.</li><li>Are water soluble, so can be difficult to reclaim from a liquid.</li></ul>

12.7

How do our bodies keep internal conditions constant?

The body's internal environment stays constant by removing waste products regularly. To understand this, you need to know...

- what waste products need to be removed, and how they are removed
- which internal body conditions need to be controlled and how the body controls them
- what happens when the body's water or ion content is wrong
- what happens when the core temperature is too high or too low
- what diabetes is and what causes it.

Controlling Conditions

The water content, ion content, temperature and blood sugar levels of the human body have to be controlled so that it can function properly.

Water and Ion Content

If the water or ion content of the body is wrong, too much water may move into or out of the cells by osmosis, causing damage. Water and ions enter the body through food and drink.

Blood Glucose Concentration

Blood glucose concentration is monitored and controlled by the pancreas which secretes the hormone insulin.

Insulin converts glucose into insoluble glycogen and lowers blood glucose (i.e. insulin allows glucose to move from the blood into the cells).

The level of insulin in the pancreas affects what happens in the liver. The pancreas continually monitors the body's blood sugar levels and adjusts the amount of insulin released to keep the body's blood sugar levels as close to normal as possible.

If the blood glucose concentration is too high...

the pancreas releases insulin.

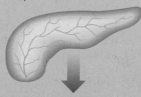

Glucose from the blood is then converted to insoluble glycogen in the liver...

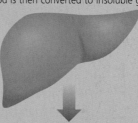

and removed from the blood.

The blood glucose concentration returns to normal.

If the pancreas does not produce enough insulin, a person's blood glucose concentration may rise to a fatally high level. This is a condition called **diabetes**, which is treated by...

- careful attention to diet
- injecting insulin into the blood.

Unit 2

Body Temperature

Body temperature should be kept at around 37°C because this is the ideal temperature for enzymes. It is controlled by the nervous system.

Monitoring and control is performed by the thermoregulatory centre in the brain, which has receptors that are sensitive to the temperature of blood flowing through it. There are also temperature receptors in the skin, which provide information about skin temperature. Sweating helps to cool the body. More water is lost when it is hot, and more water has to be taken in as food or drink to balance this.

Removing Waste Products

Humans need to remove waste products from their body to keep their internal environment relatively constant. Two of these waste products are carbon dioxide and urea.

- Carbon dioxide is produced by respiration and is removed via the lungs during exhalation.
- Urea is produced by the liver when it breaks down amino acids, and is removed by the kidneys and transferred to the bladder before it is released.

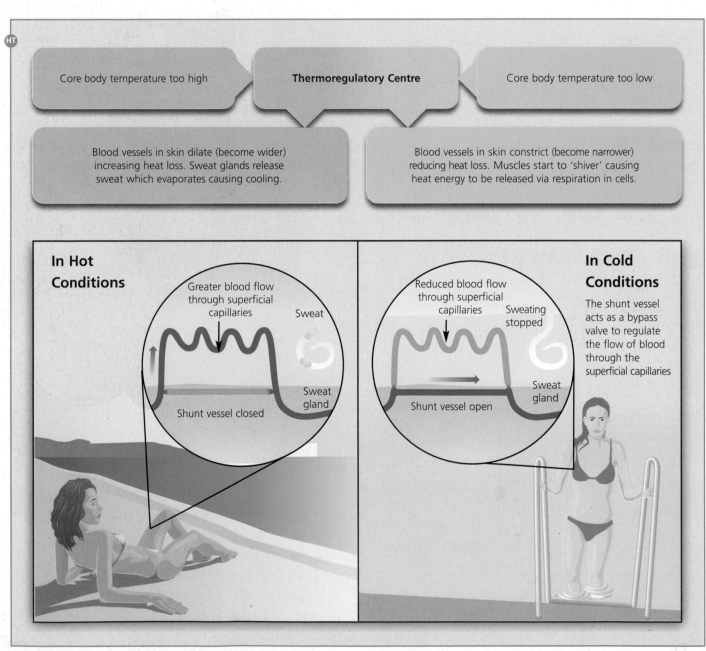

How Science Works

You need to be able to evaluate the data from the experiments by Banting and Best which led to the discovery of insulin.

In the early 20th century, it was discovered that the pancreas played a part in diabetes.

Shortly after this discovery, Frederick Banting and Charles Best worked in the University of Toronto (in Canada) conducting experiments on dogs to discover what would happen when they removed their pancreases. As they expected, the dogs developed diabetes.

Banting and Best then developed treatments to test on the dogs. They extracted many compounds from the islets of Langerhans (cells in the pancreas which produce insulin). These compounds were then injected into the diabetic dogs to try to find the hormone that would reverse their diabetes. New methods of testing blood sugar levels allowed them to accurately determine what effects their treatments were having on the dogs.

At the start, the injections were not pure and often the dog died. With the help of a group of researchers, they were eventually able to make an extract that was pure enough to try on a human patient.

In 1922, they announced that they had discovered insulin and a 14-year-old boy was successfully treated for diabetes in Toronto Hospital with the extract that they called insulin.

Even without the data from their experiments, it is possible to evaluate how accurate and reliable the data they obtained would have been by looking at the methods they used.

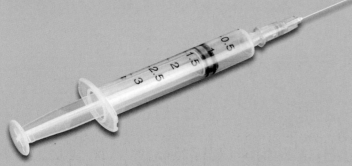

Advantages of their Methods

- They repeated their experiments and obtained the same results, which shows that the evidence is reliable.
- They carried out all their experiments on the same animal species.

Disadvantages of their Methods

- All their experiments were carried out on dogs; there was no guarantee that the same results would be produced in humans.
- Initially, they could not control the purity of the extract they produced.
- The equipment they were using to measure blood sugar is unlikely to have been as accurate as modern technology.
- They had only treated one person successfully with the insulin when they announced they had found a treatment – it could have been due to other factors.

How Science Works

You need to be able to evaluate modern methods of treating diabetes.

Example

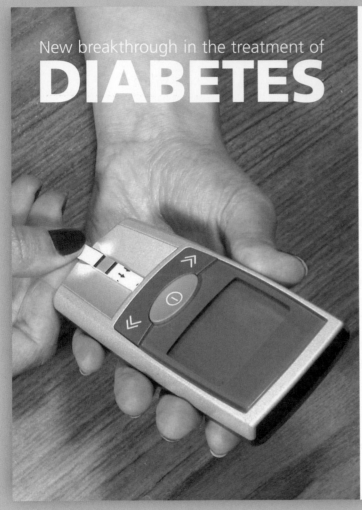

New breakthrough in the treatment of
DIABETES

Diabetes affects about 2 million people in the UK, including up to 1 million who are unaware that they are diabetic.

There are two types of diabetes. Type 1 diabetes develops suddenly, usually at a young age. People with Type 1 diabetes need daily injections of insulin throughout their lives since their bodies do not produce enough insulin. They also require a healthy diet with the right balance of foods. Type 2 diabetes develops gradually and tends to affect people over 40 who are often overweight or obese. Type 2 diabetes can usually be controlled through diet and exercise alone but, in some cases, people with Type 2 diabetes may also need insulin injections.

Although with practice most people find giving themselves injections pain-free and simple, it can be difficult for those who need up to four insulin injections a day. Further difficulties are often encountered with injecting, as the absorption of insulin into the bloodstream can be affected by a number of factors. These include the site of the injection (insulin is absorbed quickest from the abdomen), and using the correct injection technique (if it is too shallow into the skin, or too deep into the muscle, the insulin will not be absorbed properly). Smoking can cause inconsistent blood glucose levels as nicotine causes changes in small blood vessels which may affect absorption.

However, a new breakthrough treatment is being developed, which would allow powdered insulin to be taken through an inhaler. This could be a fantastic step forward for all insulin users, and would reduce problems associated with injections.

Research carried out on inhaled insulin found that the doses produced regular concentrations of insulin. However, there are problems too, for example, the timing of doses would be difficult to assess as the response to a dose may not be easy to predict. Inhaling insulin may prove to be expensive, due to the large doses required, whilst another concern is that it may have the potential to cause lung cancer.

Injected Insulin

Benefits

- Treatment has been used successfully for many years.
- Most people find injecting themselves painless and fairly easy.

Problems

- People must learn the correct way to inject, which can be difficult.
- Some people need to inject themselves up to four times a day, which can be inconvenient.
- Absorption into the bloodstream can be affected by the site of the injection.
- Further problems can be encountered if the person smokes.

Inhaled Insulin

Benefits

- People don't have to inject themselves.
- May lead to the development of other methods of administering insulin.
- In tests, the inhaled doses produced consistent concentrations of insulin.

Problems

- Research is still being carried out, so it has not been widely used yet.
- Timing of doses would be difficult to work out.
- It may be expensive.
- It may have the potential to cause lung cancer.

12.8

Which human characteristics show a simple pattern of inheritance?

Certain genetic disorders and the sex of a baby show a simple pattern of inheritance. To understand this, you need to know…

- the difference between body cells and sex cells in terms of chromosomes
- what meiosis and mitosis are
- why sexual reproduction promotes variation
- how genes control sex and characteristics
- what alleles, chromosomes and DNA are
- how certain disorders are inherited, and how embryo screening can identify these disorders.

Body cells contain **46 chromosomes** arranged as **23 pairs**. Chromosomes are made up of large molecules of **DNA**. A **gene** is a section of DNA.

Gametes – female eggs and male sperm – have 23 chromosomes (one from each pair). The fusion of these two cells produces a **zygote** with 46 chromosomes in total (23 pairs).

Inheritance of Sex – the Sex Chromosome

Of the 23 pairs of chromosomes in human body cells, one pair is the sex chromosomes.

- In females, these are identical and are called the X chromosomes.
- In males, one is much shorter than the other. The shorter one is called the Y chromosome and the longer one is called the X chromosome.

Female Sex Chromosomes   **Male Sex Chromosomes**

Like all pairs of chromosomes, offspring inherit one sex chromosome from the mother and one from the father.

Ultimately, therefore, the sex of an individual is decided by whether the ovum is fertilised by an X-carrying sperm or a Y-carrying sperm.

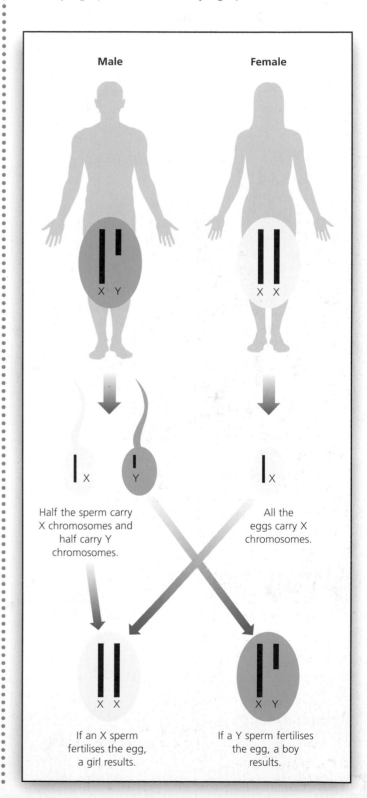

Male   Female

X Y   X X

X   Y   X

Half the sperm carry X chromosomes and half carry Y chromosomes.   All the eggs carry X chromosomes.

X X   X Y

If an X sperm fertilises the egg, a girl results.   If a Y sperm fertilises the egg, a boy results.

Cell Division

Mitosis

Mitosis is the division of body cells to produce new cells. This occurs for growth and repair (and also in asexual reproduction). Before the cell divides, a copy of each chromosome is made so the new cell has exactly the same genetic information. This means that the cells of asexually reproduced offspring contain the same genes as the parents.

Fertilisation

When gametes join at fertilisation, one chromosome comes from each parent and a single body cell with new pairs of chromosomes is formed.

This then divides repeatedly by mitosis to form a new individual, giving rise to variation.

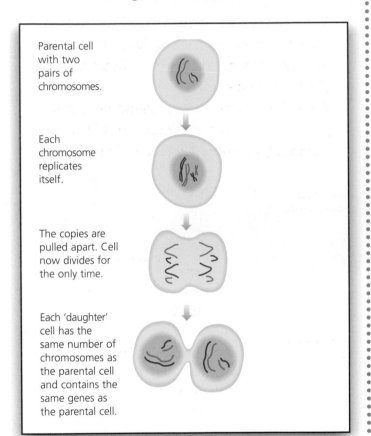

Parental cell with two pairs of chromosomes.

Each chromosome replicates itself.

The copies are pulled apart. Cell now divides for the only time.

Each 'daughter' cell has the same number of chromosomes as the parental cell and contains the same genes as the parental cell.

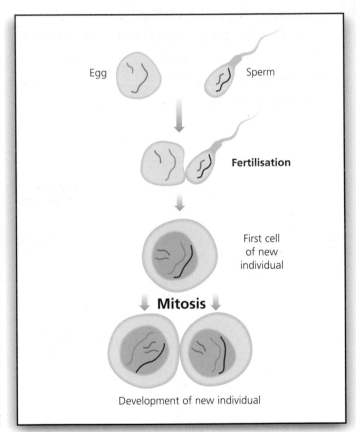

Egg   Sperm

Fertilisation

First cell of new individual

Mitosis

Development of new individual

HT ### Meiosis

This occurs in the testes and ovaries. The cells in these organs divide to produce the gametes (eggs and sperm) for sexual reproduction.

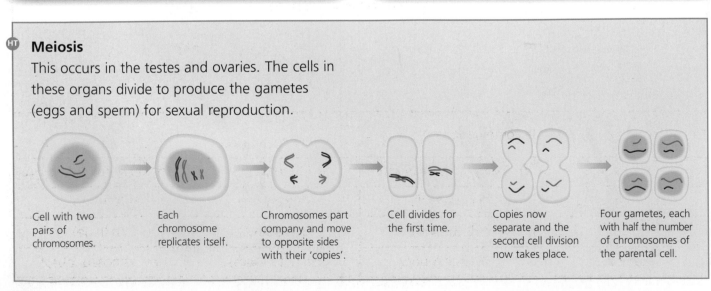

Cell with two pairs of chromosomes.

Each chromosome replicates itself.

Chromosomes part company and move to opposite sides with their 'copies'.

Cell divides for the first time.

Copies now separate and the second cell division now takes place.

Four gametes, each with half the number of chromosomes of the parental cell.

Genetics

Alleles

Some genes have different forms or variations – these are called **alleles**. For example, the gene that controls whether you can roll your tongue or not has two alleles (or forms) – you either can or you can't. Similarly, in the gene for eye colour there are two alleles – blue or brown.

In a pair of chromosomes, the alleles for a gene can be the same or different. If they are different, then one allele will be **dominant** and one allele will be **recessive**. The dominant allele will control the characteristics of the gene. A recessive allele will only control the characteristics of the gene if it is present on both chromosomes in a pair (i.e. no dominant allele is present).

Example: Dominant and Recessive Alleles

Here are three pairs of genes from the middle of a pair of chromosomes. These are the genes which code for tongue-rolling ability, eye colour, and type of earlobe.

- **Dominant alleles** express themselves even if present only once so an individual can be **homozygous dominant** (BB) or **heterozygous** (Bb) for brown eyes.
- **Recessive alleles** express themselves only if present twice so an individual can only be **homozygous recessive** (bb) for blue eyes.

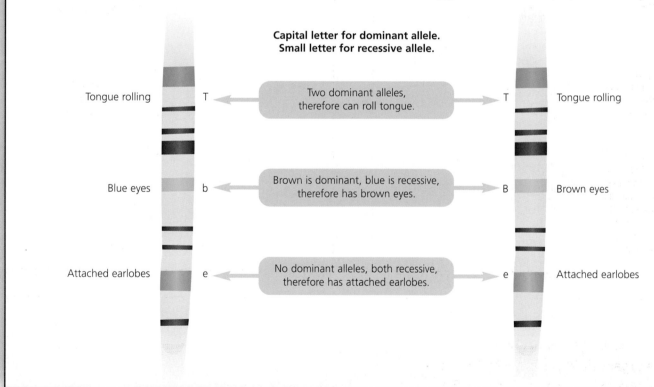

Capital letter for dominant allele.
Small letter for recessive allele.

Tongue rolling — T — Two dominant alleles, therefore can roll tongue. — T — Tongue rolling

Blue eyes — b — Brown is dominant, blue is recessive, therefore has brown eyes. — B — Brown eyes

Attached earlobes — e — No dominant alleles, both recessive, therefore has attached earlobes. — e — Attached earlobes

---

This table shows the possible combinations:

	Homozygous Dominant	**Heterozygous**	**Homozygous Recessive**
Tongue rolling	TT (can roll)	Tt (can roll)	tt (can't roll)
Eye colour	BB (brown)	Bb (brown)	bb (blue)
Ear lobes	EE (free lobes)	Ee (free lobes)	ee (attached lobes)

Monohybrid Inheritance

As we saw on previous pages, genes exist in pairs; one on each chromosome in a pair. We call these pairs of genes 'alleles' when they code for alternatives of the same characteristic, e.g. eye colour. When a characteristic is determined by just one pair of alleles then simple genetic crosses can be performed to investigate the mechanism of inheritance. This type of inheritance is referred to as **monohybrid inheritance**.

Inheritance of Eye Colour

In genetic diagrams we use capital letters for dominant alleles and lower case for recessive alleles. In eye colour, therefore, we use 'B' for brown eye alleles and 'b' for blue eye alleles.

From the crosses on the diagrams opposite it can be seen that…

1. if one parent has two dominant genes then all the offspring will inherit that characteristic

2. if both parents have one recessive gene then this characteristic may appear in the offspring (a one in four chance)

3. if one parent has a recessive gene and the other has two recessive genes, then there is a 50% chance of that characteristic appearing.

But remember, these are only probabilities. In practice, all that matters is which egg is fertilised by which sperm, and that is completely random.

HT These are the typical examples you may be asked about in your exam. When you construct genetic diagrams remember to…
- clearly identify the alleles of the parents
- place each of these alleles in a separate gamete
- join each gamete with the two gametes from the other parent.

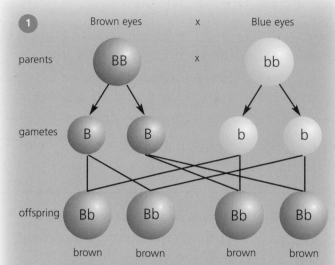

100% chance of offspring having brown eyes.

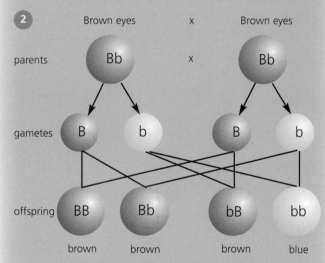

75% chance of offspring having brown eyes, 25% chance of offspring having blue eyes.

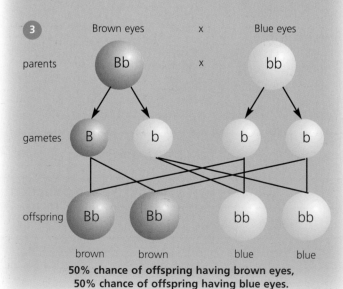

50% chance of offspring having brown eyes, 50% chance of offspring having blue eyes.

Differentiation of Cells

When cells develop a specialised structure to carry out a specific function, this is **differentiation**.

- Most plant cells can differentiate throughout life.
- Animal cells differentiate at an early stage so quickly become muscle, nerves, etc.

Mature cells usually divide for repair and replacement.

Stem Cells

Stem cells are cells in human embryos and adult bone marrow, which have yet to differentiate. They can be made to differentiate into many different types of cells, e.g. a nerve cell, so treatment with these cells may help conditions such as paralysis.

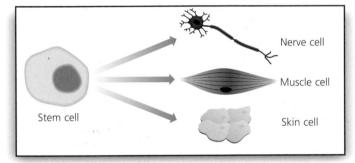

Genes, Chromosomes and DNA

In normal human cells, there are only 23 pairs of chromosomes. They consist of long, coiled molecules of **DNA** (Deoxyribonucleic acid).

The DNA molecule itself consists of two strands which are coiled to form a **double helix**.

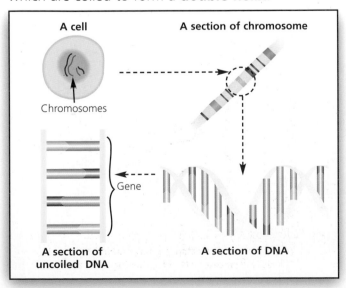

Genes are sections of DNA which code for a particular inherited characteristic, e.g. blue eyes.

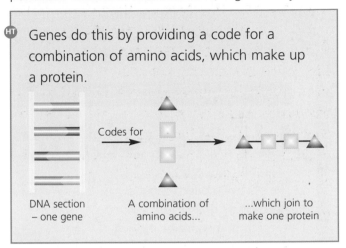

HT Genes do this by providing a code for a combination of amino acids, which make up a protein.

DNA section – one gene | A combination of amino acids... | ...which join to make one protein

Each person has unique DNA (apart from identical twins who have the same DNA), which means it can be used for DNA fingerprinting (identification).

Genetic Disorders

Embryos can be screened for the genes that cause genetic disorders.

- **Huntington's Disease**
  Huntington's disease is a disorder of the nervous system caused by a dominant allele. It can be passed on by only one parent who has the disorder.
- **Cystic Fibrosis**
  Cystic fibrosis is a disorder of cell membranes. It is caused by a recessive allele, so it must be inherited from both parents, who can be carriers without having the disorder themselves.

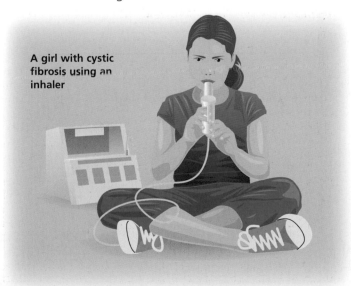

A girl with cystic fibrosis using an inhaler

How Science Works

You need to be able to explain why Mendel proposed the idea of separately inherited factors and why the importance of this discovery was not recognised until after his death.

Science Monthly

Edition 2

Understanding Genetics

The start of modern genetics was marked by Gregor Mendel's research on pea plants in 1865. He investigated the height of pea plants which were all either tall or dwarf.

He began by taking a plant which was pure-breeding for tallness (i.e. when bred with itself or other tall plants they only produced tall plants) and a plant which was pure-breeding for dwarfness (see diagram 1 below).

He cross-fertilised these plants by taking pollen from each one. To his surprise, all the plants produced from the cross-fertilisation were tall (see diagram 2 below), which led to Mendel's first law:

'When pure-breeding plants with contrasting traits are cross-fertilised, all the offspring will resemble one of the parents.'

Mendel then crossed several of the tall plants that he had produced. Again, he was surprised to find that there were three tall plants to every dwarf plant (see diagram 3 below). He based his second law on this, stating:

'For every trait, every individual must have two determiners.'

Mendel's work was not recognised in his lifetime because much of his research was carried out in his spare time. He worked alone with no one to assist him and when he sent the results of his research to a renowned German botanist he was advised that he needed more data. His results came from over 21 000 plants!

His work was discovered in 1900, 16 years after his death, by a team of scientists looking for evidence to support Charles Darwin's *'Origin of the Species'*. They considered Mendel's work to be of great importance because…

- he had planned his experiments on a large scale to eliminate chance effects

- his choice of plants was ideal since the peas showed definite characteristics and were not susceptible to disease and weather factors

- he used only pure-breeding plants to start his work.

The 'determiners' that Mendel had written about are what are now known as genes. The scientists coined the term 'genetics' and built on Mendel's work.

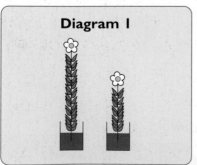

Diagram 1

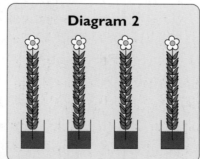

Diagram 2

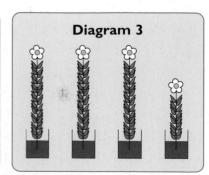

Diagram 3

12

You need to be able to make informed judgements about the social and ethical issues concerning the use of stem cells from embryos in medical research and treatments.

Most adult cells in the body have a specific function and structure which cannot be changed, for example, muscle cells cannot become nerve cells.

Stem cells are different. They are still at an early stage of development, which means they have not yet developed a special structure and function. Scientists are currently carrying out research which should eventually make it possible to use stem cells to generate healthy tissue to replace tissue that has been damaged, either by trauma or disease. Some of the conditions which scientists believe may eventually be treated by stem cell therapy are Parkinson's disease, Alzheimer's disease, heart disease, stroke, arthritis, diabetes, burns and spinal cord damage.

There is great debate about the use of stem cells; you have to form your own opinion based on evidence and understand that science can help us in many ways but it cannot supply all the answers.

To help you to make an informed decision, the table below contains some reasons for and against the use of stem cells.

A magnified stem cell.

For	Against
<ul><li>The embryos from which the stem cells are taken are grown in laboratories and are only a few days old; many people see them simply as microscopic balls of cells.</li><li>Embryos provide the most useful stem cells: the cells are non-specialised, which means they can become any specific type of cell (adult stem cells are more limited in their potential uses).</li><li>They can act as a repair kit for damaged tissue.</li><li>Stem cells may be able to treat many diseases and conditions in the future.</li></ul>	<ul><li>Many people believe it is morally wrong to experiment on embryos (even those grown in the laboratory) as they could all potentially develop into a baby.</li><li>Stem cells are cultivated using nutrients from animal sources, which could carry diseases that could be passed to humans.</li><li>People who receive cell transplants through stem cell therapy could be infected with viruses.</li><li>Stem cells may turn cancerous.</li></ul>

How Science Works

You need to be able to make informed judgements about the economic, social and ethical issues concerning embryo screening.

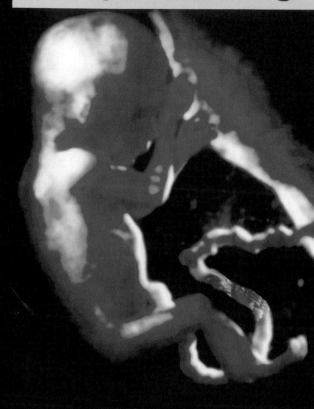

Embryo Screening

Embryos can be screened during the course of pregnancy. Screening can also take place at the eight-cell stage of development of an IVF (*in vitro* fertilisation) embryo before it is implanted into the mother's womb. This type of screening is called Pre-implantation Genetic Diagnosis (PGD) and is currently permitted in order to detect inherited diseases such as Huntington's disease and cystic fibrosis.

The Human Fertilisation and Embryology Authority is now asking if embryos should also be checked for genes linked to cancer and the early onset of Alzheimer's disease. Carrying these genes does not mean a person will definitely develop disease, but it puts them at an increased risk.

A mother who lost an eye as a child because she had retinoblastoma (eye cancer) passed on the gene to her first son, who underwent chemotherapy to get rid of the cancer. Now fertility experts at a London hospital have been given authorisation to test her embryos to find out if any of her future children are at risk of contracting the disease.

There is a fear that this type of screening could be used to detect other genetic factors, which could lead to a 'designer baby' culture, where babies who do not have the desired attributes could be aborted.

When making judgements on embryo screening, you should consider social, ethical and economic issues. Some are listed in the table below.

Advantages	Disadvantages
• Doctors can determine whether a child will have an increased risk of contracting a particular illness or disease. • It prepares parents for the possibility of a child developing a disorder, disease or illness. • Carriers of genetic disorders could make informed decisions about whether to have children.	• Could result in unborn children being aborted if their genetic make-up is 'faulty'. • Parents may want to choose the genetic make-up of their child. • It has the potential to be used to determine who can / cannot reproduce. • It could stigmatise and upset people to learn they carry a genetic disorder, disease or illness.

(HT) **You need to be able to predict / explain the outcome of crosses between individuals for each possible combination of dominant and recessive alleles of the same gene, and construct genetic diagrams.**

Example

If a gene for retinoblastoma (a type of cancer affecting the eye) is present in an individual, there is a 90% chance that they will develop the disease.

To find out the likelihood of a child receiving this gene, we can construct a simple genetic diagram.

When constructing genetic diagrams, it is important to use the upper case and lower case of the same letter, so you can easily see which is the dominant gene. The gene that causes retinoblastoma is dominant so we will call this 'R' and the normal gene which does not cause the disease, 'r'. These are alleles since they control the same characteristic and are found in pairs. The possible combinations for these alleles are…

RR – child has the retinoblastoma gene
Rr – child has the retinoblastoma gene
rr – child does not have the retinoblastoma gene.

In this example, the father does not have the disease so his gene pair must be rr. The mother has the disease so her gene pair must be RR or Rr.

To calculate the chance of an embryo carrying the gene for retinoblastoma we need to draw two genetic diagrams to show all the possibilities (see opposite).

Remember, even if an individual does have the gene for retinoblastoma, there is still a 10% chance that they will not develop the disease. And experts say nine out of ten cases of the disease can be cured. This raises questions about the ethics of using technology to genetically screen embryos (see p.76).

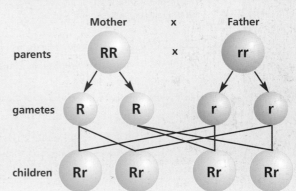

All of the crosses have an 'R' gene, so there is a 100% chance that a child will have the retinoblastoma gene.

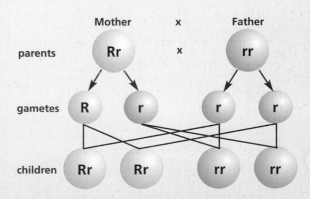

Half of the crosses produce 'Rr' and half produce 'rr', which means that there is a 50% chance that a child will have the retinoblastoma gene.

Example Questions

For Unit 2, you will have to complete one written paper with structured questions.

1 An investigation was carried out to see what effect varying the water content in the soil would have on the height of five tomato plants.

The following results were obtained.

Water content of soil	Height of tomato plants (cm)
Dry	40
Damp	55
Fairly wet	75
Very wet	75
Flooded	60

(a) What was the independent variable in the test?

The water content in the soil.

(1 mark)

(b) What was the dependent variable in the test?

The height of the tomato plants.

(1 mark)

(c) What conclusion can you draw from the investigation?

Fairly wet or very wet soil produces the tallest tomato plants.

(1 mark)

(d) Identify one problem with the method of the investigation.

Only one tomato plant was used in each condition OR There was no accurate measure of the water content in the soil, so the investigation could not be repeated OR We do not know if the tomato plants were the same height or age at the beginning of the investigation.

(1 mark)

(e) Suggest one way in which you could improve the reliability of the results.

Use more plants in each condition OR Measure the amount of water put into the soil, using a measuring cylinder OR Use plants that are all the same height to start off with OR measure the increase in height for each plant.

(1 mark)

1 Read the information carefully to make sure you understand what the data in a table shows before answering any questions.

2 The independent variable is the one that is being controlled.

3 The dependent variable is the one that is measured each time the independent variable is changed.

4 Describe what the data tells you – nothing else!

5 You only need to give **one** answer here, but there are lots of possible problems. Think about what you would need to do to improve the reliability and accuracy of the data.

6 This question relates to part (d). Think about how the problem you mentioned could be minimised or solved.

Key Words

Aerobic respiration – respiration using oxygen which releases energy and produces CO_2 and water

Allele – an alternative form of a particular gene

Amylase – an enzyme that breaks down starch

Anaerobic respiration – incomplete breakdown of glucose without oxygen to produce a small amount of energy very quickly

Bile – a greenish-yellow fluid produced by the liver

Biomass – the mass of a plant or animal minus the water content

Bone marrow – soft, spongy tissue in bones

Capillary – the narrowest type of blood vessel

Catalyst – a substance that increases the rate of a chemical reaction without being changed itself

Cell – a fundamental unit of a living organism

Chlorophyll – the green pigment found in most plants, responsible for photosynthesis

Chloroplasts – tiny structures in the cytoplasm of plant cells which contain chlorophyll

Community – the total collection of living organisms within a defined area or habitat

Constrict – to compress; make narrow or tight

Core temperature – the main operating temperature of an organism, in comparison to temperatures of peripheral tissues

Cytoplasm – the protoplasm of a living cell which is found outside the nucleus

Decay – to rot or decompose

Detritus – organic material formed from dead and decomposing plants and animals

Differentiate – to make / become different

Diffusion – the moving of particles from high to low concentration

DNA – nucleic acid which contains the genetic information carried by every cell

Environment – the conditions around an organism

Enzyme – a protein catalyst which alters the rate of a particular biochemical reaction

Fertilisation – the fusion of the male nucleus with the female nucleus

Food chain – the feeding relationship between organisms in an ecosystem

Gamete – a specialised sex cell formed by meiosis

Insoluble – a substance that will not dissolve in water

Limiting factors – the main factors that can affect the rate, e.g. of photosynthesis: light intensity; carbon dioxide concentration; temperature

Lipase – an enzyme which breaks down fat into fatty acids and glycerol

Meiosis – cell division that forms daughter cells with half the chromosome number of the parent cell

Microorganism – very small organisms

Mitochondria – the structure in the cytoplasm where energy is produced from chemical reactions

Mitosis – cell division that forms two daughter cells, each with the same number of chromosomes as the parent cell

Neutralise – to make the pH level neutral (pH7)

Nitrate – any compound containing nitrogen

Nucleus – the control centre of a cell

Organ – a collection of tissues which work together to perform a function in the organism, e.g. heart

Osmosis – the movement of water from a dilute to a more concentrated solution across a selectively permeable membrane

Permeable – allows a substance to pass through

pH – measure of the strength of an acid or alkali

Photosynthesis – the chemical process where water combines with carbon dioxide to produce glucose using light energy

Protease – an enzyme used to break down proteins into amino acids

Ribosomes – particles involved in the synthesis of proteins

Soluble – a substance that dissolves in water to form a solution

Specialised – developed for a special function

Thermoregulation – maintenance of a constant body temperature in warm-blooded animals

Tissue – a collection of similar cells which perform a specific function, e.g. skin tissue

Urea – toxin produced when proteins are broken down

Urine – waste product produced by kidneys

Vacuole – a fluid-filled sac found in cytoplasm

13.1

How do dissolved materials get into and out of animals and plants?

Animal and plant cells need oxygen to function, and produce carbon dioxide as a waste product. To understand this, you need to know…

- how substances are transported
- the ways in which organs are adapted to exchange materials
- how substances are exchanged in humans and plants.

Water and dissolved substances automatically move along a concentration gradient, from high concentrations to low concentrations, by osmosis and diffusion (see p.52).

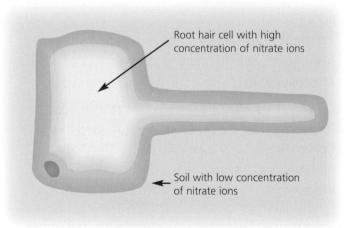

Root hair cell with high concentration of nitrate ions

Soil with low concentration of nitrate ions

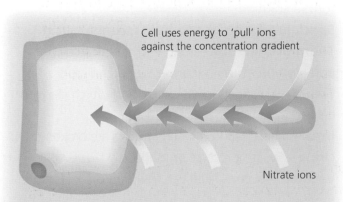

Cell uses energy to 'pull' ions against the concentration gradient

Nitrate ions

Active Transport

Substances are sometimes absorbed **against** a concentration gradient. This requires the use of energy from respiration and is known as **active transport**. Plants absorb ions from very dilute solutions in this way (see diagram opposite).

Active transport takes place in the opposite direction to which normal diffusion would occur.

This process of active transport requires energy (from respiration) in just the same way that pushing a ball up a hill would require energy.

Sugar and ions, which can pass through cell membranes, can also be moved by active transport.

In humans, sugar may be absorbed from the intestine and from the kidney tubules by active transport.

Exchanging Material in Humans

In humans there are organ systems that are specialised to aid the exchange of materials.

Villi in the Small Intestine

Villi line the walls of the small intestine. They have a massive surface area and an extensive network of capillaries.

This network absorbs the products of digestion by diffusion and active transport.

A single villus

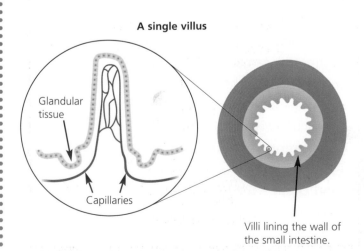

Glandular tissue

Capillaries

Villi lining the wall of the small intestine.

Exchanging Material in Humans (cont.)

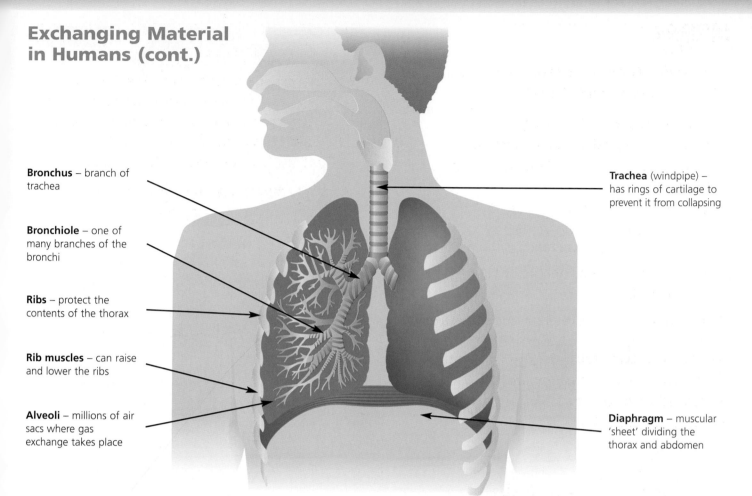

Bronchus – branch of trachea

Bronchiole – one of many branches of the bronchi

Ribs – protect the contents of the thorax

Rib muscles – can raise and lower the ribs

Alveoli – millions of air sacs where gas exchange takes place

Trachea (windpipe) – has rings of cartilage to prevent it from collapsing

Diaphragm – muscular 'sheet' dividing the thorax and abdomen

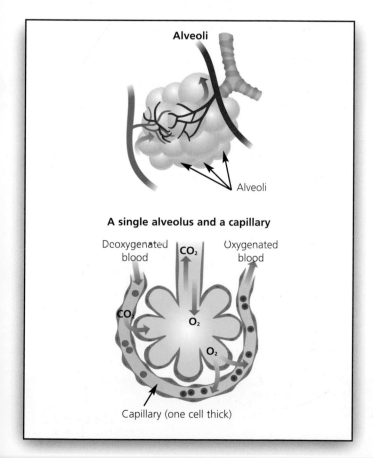

Alveoli

Alveoli

A single alveolus and a capillary

Deoxygenated blood

CO_2

Oxygenated blood

CO_2

O_2

O_2

Capillary (one cell thick)

Alveoli in the Lungs

The diagram above shows the breathing system. The ribcage protects the contents of the thorax, i.e. the heart and lungs. The breathing system takes air into and out of the body.

The trachea divides into two tubes called the **bronchi**, which divide again, several times, to form the **bronchioles** which continue to divide until they end in **air sacs** called **alveoli** (there are millions of these), which are very close to the blood capillaries.

The arrangement of alveoli and capillaries in the lungs makes them very efficient at exchanging oxygen and carbon dioxide, because…
• they have a large, moist surface area
• they have an excellent blood supply.

Carbon dioxide diffuses from the blood into the alveoli, and oxygen diffuses from the alveoli into the blood. This means the blood has swapped its carbon dioxide for oxygen and is now oxygenated.

Exchange in Plants

The Leaf

Leaves are broad, thin and flat with lots of internal air spaces, providing a large surface area in order to make them efficient at photosynthesis. They also have **stomata** on their undersurface to allow carbon dioxide in and oxygen out (by diffusion). This, however, also leads to loss of water vapour in a process called **transpiration**.

Water loss through transpiration is the price the plant must pay in order to photosynthesise. Transpiration is quicker in hot, dry, windy conditions.

Water vapour from the internal leaf cells evaporates through the stomata. However, the size of the stomata is controlled by a pair of guard cells. If plants lose water faster than it is taken up by the root hair cells in the roots, the stomata can close to prevent wilting and eventual dehydration.

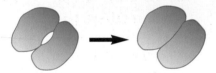

In periods of intense drought, photosynthesis may be impossible since the stomata will be closed to prevent water loss.

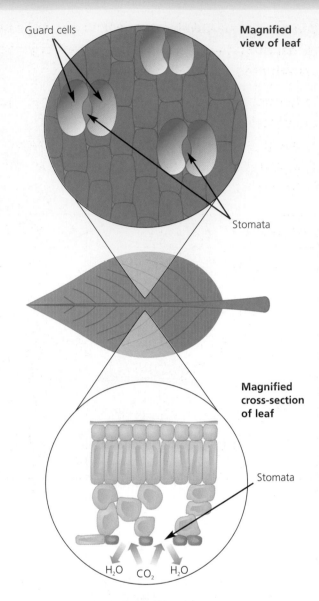

Guard cells

Magnified view of leaf

Stomata

Magnified cross-section of leaf

Stomata

H_2O CO_2 H_2O

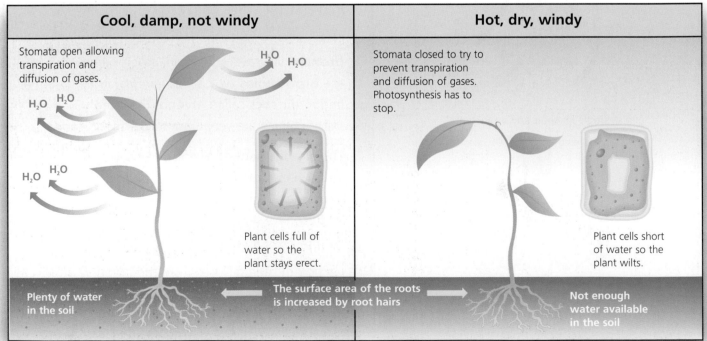

Cool, damp, not windy	Hot, dry, windy
Stomata open allowing transpiration and diffusion of gases.	Stomata closed to try to prevent transpiration and diffusion of gases. Photosynthesis has to stop.

H_2O H_2O

H_2O H_2O

H_2O H_2O

Plant cells full of water so the plant stays erect.

Plant cells short of water so the plant wilts.

Plenty of water in the soil

The surface area of the roots is increased by root hairs

Not enough water available in the soil

You need to be able to explain how gas and solute exchange surfaces in humans and other organisms are adapted to maximise effectiveness.

To function efficiently, exchange surfaces need the following features:

- a large surface area to maximise the exchange
- a method of transporting substances to and from the exchange site.

In small organisms, the 'skin' or outer membrane functions as an exchange surface, allowing gases and solutes into and out of the body by diffusion.

Example

The table below looks at how different small organisms facilitate the exchange of oxygen and carbon dioxide (aerobic respiration).

Organism	Gas Exchange System	Explanation
Earthworms	• Entire surface of body. gut skin	The earthworm has adapted so that its whole surface area is a gas exchange surface. Its gut extends throughout its length and is covered by skin. The skin provides the gas exchange surface. It is ideal because the skin is not very thick. The earthworm is very small and fairly inactive so it does not need much oxygen.
Flatworms	• Surface of flattened body. gut	The flatworm is a flattened version of the earthworm. Again, its whole surface area is a gas exchange surface. Its flattened shape means the materials do not have as far to diffuse. Like the earthworm they are often inactive, so they do not require much oxygen.
Lugworms and tadpoles	• External gills (outside the body) plus rest of surface of body. Artery brings deoxygenated blood from the heart Artery takes oxygenated blood from the gills to tissues Direction of water flow Gaseous exchange surface	Lugworms and tadpoles have external gills and the surface of their body also provides a surface for gas exchange. The gills increase the surface area of the gas exchange surface. However, they can be easily damaged.

How Science Works

You need to be able to explain how gas and solute exchange surfaces in humans and other organisms are adapted to maximise effectiveness. (Continued)

In larger organisms, which are more complex and have millions of cells, there are different specialised exchange surfaces, like the villi in the small intestine and the alveoli in the lungs (see p.80–81).

Example

The table below looks at how different larger organisms facilitate the exchange of oxygen and carbon dioxide (aerobic respiration).

Organism	Gas Exchange System	Explanation
Fish	• Gills with a dense network of blood capillaries. Artery brings deoxygenated blood from the heart Artery takes oxygenated blood from the gills to tissues Direction of water flow Gaseous exchange surface	If water flows over the fish's gills in a certain direction, this allows gases to exchange on the surface area of the gills. Fish are very active so they need a lot of oxygen. This is why they have developed gills to increase their gas exchange surface area.
Insects	• Trachea tubes that end in muscles inside throat. Muscle Air Trachea Fluid Tracheole cell	The gas exchange takes place at the end of their tracheal tube, which penetrates all parts of the insect. Insects are very active so they require a lot of oxygen.
Humans	• Alveoli – air sacs in lungs. • Fine capillaries lie next to alveoli. Bronchus Bronchiole Alveoli	Humans draw air into their lungs by ventilation. The gas exchange then takes place in air sacs in their lungs, called alveoli. The oxygen diffuses from the air sacs into the fine capillaries that lie next to them. The gases dissolve in watery fluid at the exchange surface and are taken around the body. Humans are very active so they require a lot of oxygen.

13.2

How are dissolved materials transported around the body?

Substances are carried around the body by the circulation system. These substances are taken to or from the cells. To understand this, you need to know…

- what the circulation system consists of, and how it works
- what is transported in blood plasma
- the functions of red blood cells.

The Circulation System

The **circulation system** is the body's transport system. It carries blood from the heart to all the cells of the body to provide them with food and oxygen (**oxygenated blood**), and carries waste products, including carbon dioxide, away from the cells (**deoxygenated blood**). Blood is pumped to the lungs so that carbon dioxide can be exchanged for oxygen.

The system consists of the **heart**, the **blood vessels**, and the **blood**.

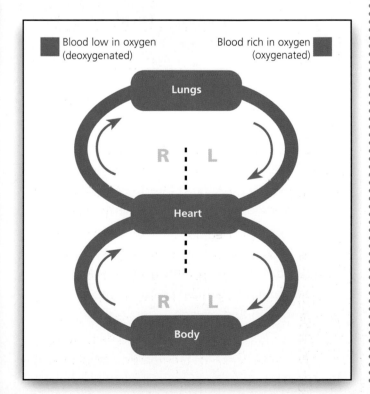

The Heart and Major Blood Vessels

The heart acts as a pump in a double circulation system. Blood flows around a 'figure of eight' circuit and passes through the heart twice on each circuit. Blood travels away from the heart through **arteries**, and returns to the heart through **veins**. In the organs, blood flows through **capillaries.** Substances needed by cells in the body tissues pass out of the blood, and substances produced by cells pass into the blood through capillary walls.

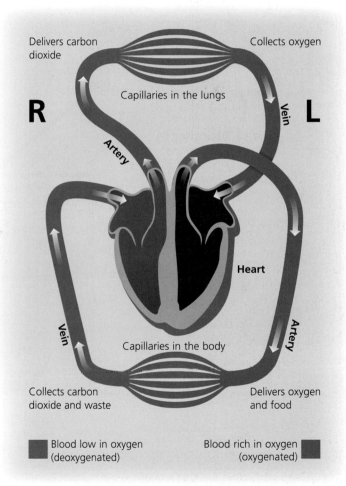

There are two separate circulation systems: one carries blood from the heart to the lungs then back to the heart, and the other carries blood from the heart to all other organs of the body then back to the heart.

- The right side of the heart pumps blood which is low in oxygen to the lungs, to pick up oxygen.
- The left side of the heart pumps blood which is rich in oxygen to all other parts of the body.

The Blood

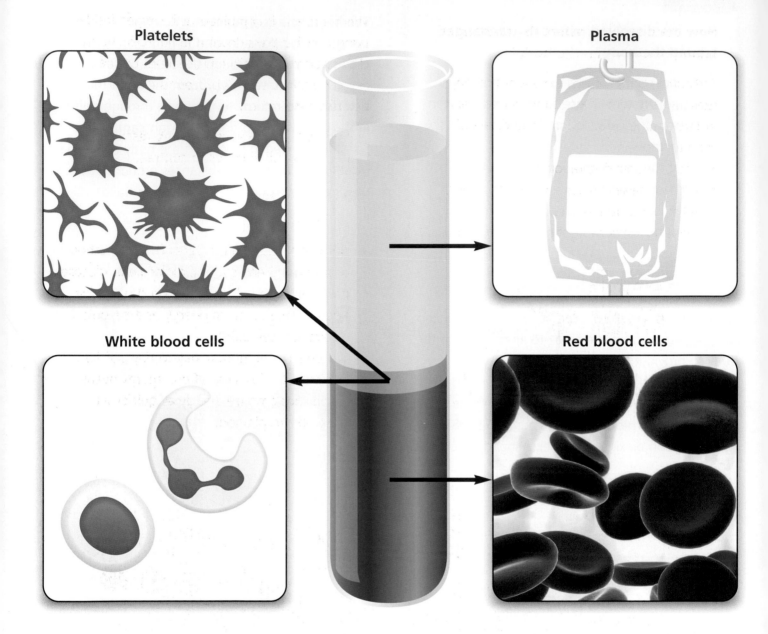

Platelets

Plasma

White blood cells

Red blood cells

If blood is allowed to stand without clotting, it separates out into its four components: plasma, red blood cells, white blood cells and platelets.

The **plasma** and **red blood cells** play an important role in the transportation of substances around the body. These are the two components you need to know about.

Plasma is a straw-coloured liquid which transports…
- carbon dioxide from the organs to the lungs
- glucose (soluble products of digestion) from the small intestine to the organs
- other wastes (e.g. urea) from the liver to the kidneys.

Red blood cells transport oxygen from the lungs to the organs.
- They have no nucleus so that they can contain lots of haemoglobin, (a red pigment which can carry oxygen).
- In the lungs, haemoglobin combines with oxygen to form **oxyhaemoglobin**. In other organs, oxyhaemoglobin splits up into haemoglobin and oxygen.

13.3

How does exercise affect the exchanges taking place within the body?

The human body requires more energy during exercise, so it must react to this demand and increase energy production. To understand this, you need to know...

- what aerobic respiration is
- what anaerobic respiration is and why it takes place during exercise
- what happens to the body during exercise.

Aerobic Respiration

When glucose is combined with oxygen inside living cells it breaks down and releases energy. (The energy is actually contained inside the glucose molecule.) This process is called **aerobic respiration**.

The energy that is released during respiration is used to enable muscles to contract.

Some Facts about Aerobic Respiration

- Aerobic respiration is a very efficient method of producing energy: one molecule of glucose produced by aerobic respiration can provide twenty times as much energy as anaerobic respiration (see p.88).
- It occurs during normal day-to-day activity and provides for most of our energy needs.
- It does not produce energy as quickly as anaerobic respiration.

A Working Muscle Cell

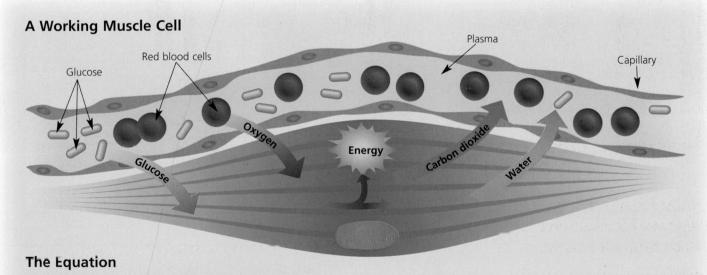

The Equation

Glucose + Oxygen ⟶ Carbon dioxide	+	Water	+	Energy
Glucose and **oxygen** are brought to the respiring cells by the bloodstream.	**Carbon dioxide** is taken by the blood to the lungs, and breathed out.	**Water** passes into the blood and is lost as sweat, moist breath and urine.	**Energy** is used for muscle contraction, metabolism and maintaining temperature.	

Anaerobic Respiration

When there is no oxygen present, glucose cannot break down completely in living cells. This incomplete breakdown of glucose releases a little bit of energy very quickly inside the cells. This is **anaerobic respiration**.

The waste product from anaerobic respiration is lactic acid which accumulates in the tissues. When this happens, the muscles become fatigued.

After exercise, the body needs oxygen to break down the lactic acid; the oxygen needed is called an **oxygen debt**.

Some Facts about Anaerobic Respiration

- Anaerobic respiration involves the incomplete breakdown of glucose. This means that much less energy is released than in aerobic respiration (about $\frac{1}{20}$).
- It can produce energy much faster than aerobic respiration over a short period of time, until fatigue sets in.
- When the muscles are fatigued and deep breathing is required to oxidise the lactic acid to carbon dioxide and water.

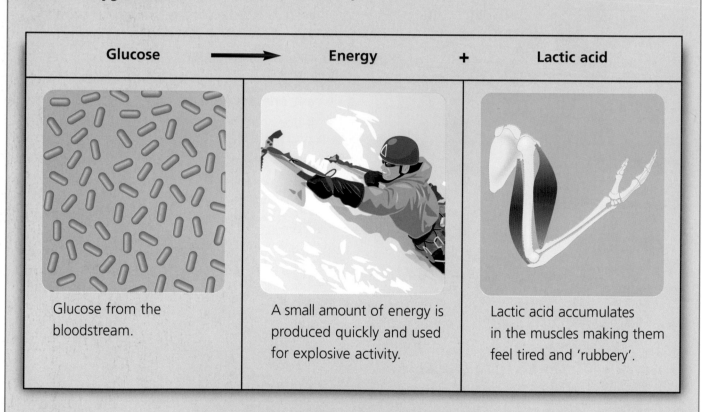

| Glucose | ⟶ | Energy | + | Lactic acid |

Glucose from the bloodstream.

A small amount of energy is produced quickly and used for explosive activity.

Lactic acid accumulates in the muscles making them feel tired and 'rubbery'.

What Happens to the Body during Exercise

During exercise a number of changes take place.
- The heart rate increases, e.g. from 70 beats per minute to 100 beats per minute.
- The arteries supplying the muscles dilate (widen).
- The rate and depth of breathing increases.

- The changes increase blood flow to the muscles.
- The supply of oxygen and sugar is increased which speeds up removal of carbon dioxide.
- Animal starch, or glycogen, stored in muscles is broken down to glucose to be used in respiration.

If muscles are subjected to long periods of vigorous activity, they become fatigued - they stop contracting efficiently.

Interpret data relating to the effects of exercise on the human body.

Athletes worldwide undertake a strict regime of exercise and diet to compete in the Olympics.

Exercise makes extreme demands on the body's energy supplies:

- glycogen stores in the muscles are used up (converted to glucose)
- more oxygen is needed by the muscles for aerobic respiration
- increased heart and breathing rates increase the supply of sugar and oxygen to the muscles.

If muscles become tired, they stop working efficiently. If insufficient oxygen reaches the muscles, they use anaerobic respiration, which can increase the recovery time after exercise. The relationship between oxygen used up during exercise and recovery can be shown in a graph.

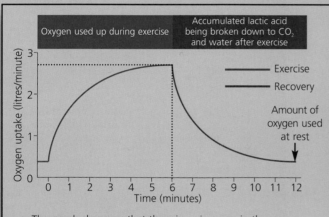

- The graph shows us that there is an increase in the oxygen uptake for 6 minutes.
- The uptake increases between 0 and 3 minutes, then stays almost constant between 5 and 6 minutes.
- After exercising for 6 minutes the recovery time is 6 minutes.
- Oxygen uptake is high at first: nearly 3 litres per minute. But it falls rapidly between 7 and 8 minutes.
- Between 9 and 12 minutes the oxygen uptake reduces until the normal resting amount of 0.3 litres per minute is reached.

It is important to remember that...

- a line of best fit can be a curve
- when answering questions about graphs, comment on each section of the graph in turn giving numbers as a guide for the examiner.

HT The length of the recovery time after exercise depends on the amount of the oxygen debt and the fitness of the athlete.

13.4

How do exchanges in the kidney help us to maintain the internal environment in mammals and how has biology helped us to treat kidney disease?

The kidneys filter the blood. If they do not work properly, toxic substances can gather in the blood. Dialysis machines can be used to help, and kidney transplants can be used to replace a failing kidney. To understand this, you need to know…

• how a healthy kidney works
• how people suffering from kidney failure can be treated.

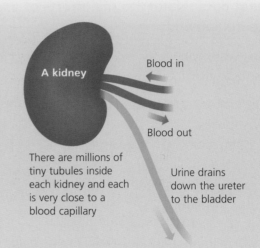

There are millions of tiny tubules inside each kidney and each is very close to a blood capillary

Urine drains down the ureter to the bladder

The Function of the Kidney

In the majority of humans there are two kidneys situated on the back wall of the abdomen. Their function is to maintain the concentrations of dissolved substances in the blood, and to remove all **urea**. If the kidneys fail then there is no way of removing excess substances. This will ultimately result in death.

Control of Ion Content and Excretion of Urea

You do not need to understand the structure of the kidney but you do need to know how it works.

It is made up of two important tissues: **blood vessels** and **tubules** (small tubes). Blood vessels take the blood through the kidney where unwanted substances end up in millions of tiny tubules, which eventually join together to form the **ureter**. The substances flow through the tubules into the ureter, which leaves the kidney and ends up at the bladder.

The kidney regulates the amount of water and ions in the blood and removes all urea.

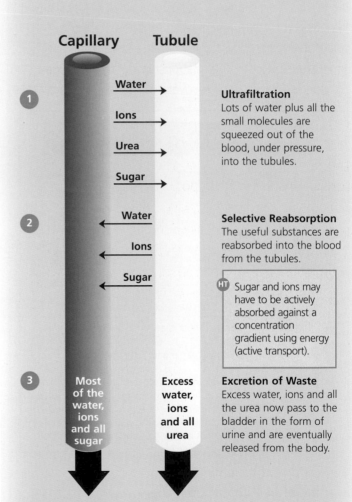

Ultrafiltration
Lots of water plus all the small molecules are squeezed out of the blood, under pressure, into the tubules.

Selective Reabsorption
The useful substances are reabsorbed into the blood from the tubules.

HT Sugar and ions may have to be actively absorbed against a concentration gradient using energy (active transport).

Excretion of Waste
Excess water, ions and all the urea now pass to the bladder in the form of urine and are eventually released from the body.

So, in principle, there are three stages to learn.
1. Nearly everything is forced out of the blood into the tubules.
2. The substances we want to keep are reabsorbed back into the blood.
3. Unwanted substances are released as **urine**.

Using a Dialysis Machine

In a dialysis machine, a person's blood flows between partially permeable membranes. These membranes are made from a material similar to the Visking tubing you may have used in experiments on osmosis.

As the blood flows through the machine, it is separated from the dialysis fluid only by the partially permeable membranes. These membranes allow all the urea, and any excess substances to pass from the blood to the dialysis fluid. This restores the concentrations of dissolved substances in the blood to their normal levels.

Dialysis fluid contains the same concentration of useful substances as blood. This ensures glucose and essential mineral ions are not lost through diffusion. Dialysis must be carried out at regular intervals in order to maintain the patient's health.

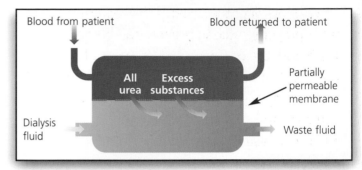

Kidney Transplants

A kidney transplant enables a diseased kidney to be replaced by a healthy one from a donor. This is performed if both kidneys fail – one kidney is sufficient to do a good job. The main problem with kidney transplants is the possibility of rejection by the immune system. The following precautions are taken to minimise the risk of rejection.

- A donor kidney with a tissue type as close as possible to that of the recipient is used. This is best achieved if the donor is a close relative.

- The bone marrow of the recipient is **irradiated** to stop the production of white cells. This reduces the likelihood of rejection just after transplantation.
- The recipient is treated with drugs which suppress the immune response. Again, this lessens the chances of rejection in the early stages.
- The recipient is kept in sterile conditions for some time after the operation to lessen the risk of infection due to his/her suppressed immune system.

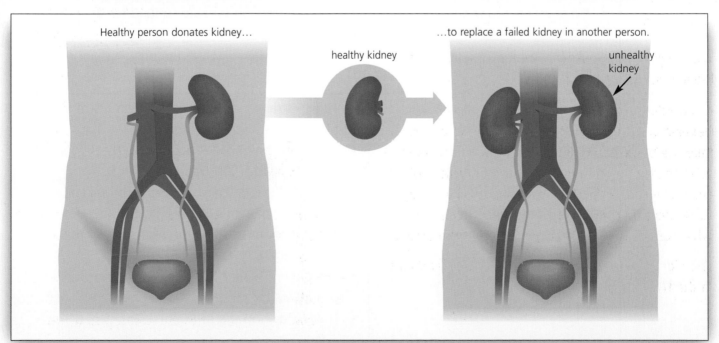

How science works

Evaluate the advantages and disadvantages of treating kidney failure by dialysis or kidney transplant.

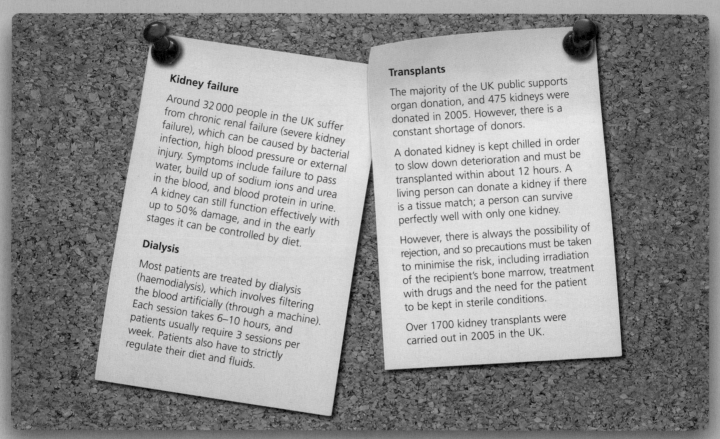

Kidney failure

Around 32 000 people in the UK suffer from chronic renal failure (severe kidney failure), which can be caused by bacterial infection, high blood pressure or external injury. Symptoms include failure to pass water, build up of sodium ions and urea in the blood, and blood protein in urine. A kidney can still function effectively with up to 50% damage, and in the early stages it can be controlled by diet.

Dialysis

Most patients are treated by dialysis (haemodialysis), which involves filtering the blood artificially (through a machine). Each session takes 6–10 hours, and patients usually require 3 sessions per week. Patients also have to strictly regulate their diet and fluids.

Transplants

The majority of the UK public supports organ donation, and 475 kidneys were donated in 2005. However, there is a constant shortage of donors.

A donated kidney is kept chilled in order to slow down deterioration and must be transplanted within about 12 hours. A living person can donate a kidney if there is a tissue match; a person can survive perfectly well with only one kidney.

However, there is always the possibility of rejection, and so precautions must be taken to minimise the risk, including irradiation of the recipient's bone marrow, treatment with drugs and the need for the patient to be kept in sterile conditions.

Over 1700 kidney transplants were carried out in 2005 in the UK.

Method	Benefits	Problems
Dialysis	• Readily available and can be used by patients waiting for a transplant. • No rejection can occur.	• Regular sessions in hospital or at home on the dialysis machine take up a lot of time. • Each session can take up to ten hours. • The diet has to be strictly regulated: some foods can only be eaten when the person is connected to the dialysis machine.
Kidney transplant	• No need for regular dialysis sessions, so the patient can lead a less restricted life. • The success rate is nearly 80% if tissue types match. • Any healthy person with a tissue match can donate a kidney and live perfectly well with only one kidney.	• Time restraints: a donated kidney must be transplanted within 12 hours. • Rejection can occur where the body's defence system 'attacks' the transplanted kidney. • Anti-rejection drugs may need to be taken for the rest of the patient's life. • The patient must undergo irradiation of the bone marrow. • The patient must be kept in hospital in sterile conditions for some time after the transplant. • There is a shortage of donors.

13.5

How are microorganisms used to make food and drink?

Microorganisms are used to make food such as bread, cheese and yoghurt, and drinks such as wine and beer. To understand this, you need to know…

- how yeast and bacteria are used in fermentation
- what is involved in the production of wine, beer and yoghurt.

Bacteria

Bacteria are used to make yoghurt and cheese. They vary in shape and have a cell wall but no distinct nucleus. They reproduce rapidly.

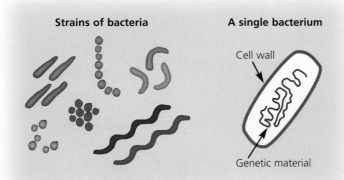

Strains of bacteria   A single bacterium

Cell wall

Genetic material

Yeast

Yeast is used to make bread and alcoholic drinks.

Yeast is a single-celled organism. Each cell has a nucleus, cytoplasm and a membrane surrounded by a cell wall.

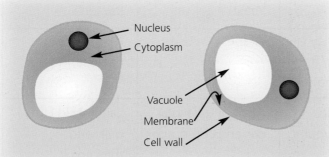

Nucleus
Cytoplasm
Vacuole
Membrane
Cell wall

How Yeast Works

Yeast can respire without oxygen (anaerobic respiration) to produce ethanol (alcohol) and carbon dioxide from glucose.

This is called **fermentation** and it has many industrial applications.

> Glucose ⟶ Ethanol + Carbon dioxide + Energy

Yeast can also respire using oxygen (aerobic respiration) to produce water and carbon dioxide.

> Glucose + Oxygen ⟶ Water + Carbon dioxide + Energy

Aerobic respiration produces more energy and is necessary for the yeast to grow and reproduce.

Bread dough is kneaded before being left in a warm place to allow the yeast to work.

Unit 3

Using Yeast in Baking

1. A mixture of yeast and sugar is added to flour.
2. The mixture is left in a warm place.
3. The carbon dioxide from the respiring yeast makes the dough rise.
4. The bubbles of gas in the dough expand when the bread is baked, making the bread 'light'.
5. As the bread is baked, any alcohol produced during respiration evaporates off.

Using Yeast in Brewing

1. In a process called malting, the starch in barley is broken down into a sugary solution by enzymes.
2. Yeast is added to the solution and fermentation takes place (see p.93). In beer-making, hops are added to give the beer flavour. In wine-making, the yeast uses the natural sugars in the grapes as its energy source.
3. Carbon dioxide is bubbled off to leave just the alcohol.

Using Bacteria to Make Yoghurt

1. A starter culture of bacteria is added to warm milk in a fermenting vessel.
2. The bacteria ferments the milk sugar (lactose) producing lactic acid which provides a sour taste.
3. The lactic acid causes the milk to clot and solidify into yoghurt.

The table below summarises the processes above.

FERMENTER

	Bread	Beer	Wine	Yoghurt
Microorganism	Yeast	Yeast	Yeast	Bacteria
Sugar Supply	Sugar is added to flour	Starch in barley is broken down into sugar (malting)	Grapes	Milk sugar (lactose)
Result	Released carbon dioxide makes bread rise	Alcohol – hops added to flavour	Alcohol – flavour depends on grapes	Lactic acid clots milk and thickens it

You need to be able to explain how scientists such as Spallanzani, Schwann and Pasteur were involved in the development of the theory of biogenesis.

Until the late 1700s most people believed in the theory of abiogenesis or 'spontaneous generation' (the process of life originating from unliving matter), which had been put forward by the famous theorist, Aristotle.

• *This is an example of how the status of the experimenter can influence the weight placed on evidence.*

In 1768 another renowned scientist, Lazzaro Spallanzani, disproved this theory by suggesting that microorganisms are found in the air. His experiments showed that microorganisms did not appear on meat broth that had been sealed in tightly closed jars and boiled for 30 minutes.

This was very forward thinking and, whilst many agreed with him, there were some who still preferred the theory put forward by Aristotle.

• *This is an example of how decisions are made by individuals and by society on issues relating to science.*

In 1839 Theodore Schwann was the first to note the importance of cells as a point from which life begins. He showed that all living things come from cells and that new cells come from existing cells. This contradicted Aristotle's theory, and formed the basis for a new theory called biogenesis.

Shortly afterwards, the famous experiments of Louis Pasteur confirmed the thinking of Spallanzani and Schwann, so that the theory of biogenesis (the theory that living things come from other living things) was confirmed.

Pasteur performed the experiment (shown below) to provide evidence for the theory.

• *This is an example of an investigation being used to produce evidence that supports a theory or hypothesis.*

Pasteur's EXPERIMENT

A solution containing nutrients was poured into a flask.

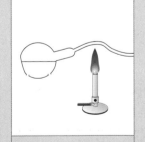

The neck of the flask was melted and pulled.

The nutrient solution was boiled to kill microorganisms and drive out air.

The nutrient solution was left for several weeks and showed no signs of decay.

When the neck was snapped off in a similar experiment, the nutrient solution started to decay in days.

Unit 3

What other useful substances can we make using microorganisms?

Substances such as penicillin (an antibiotic), mycoprotein (a food) and biogas and ethanol (fuels) can be made from microorganisms. To understand this, you need to know…

- how penicillin is made
- how mycoprotein is made
- how fuels are developed from fermentation of natural plant products and waste.

Growing Microorganisms

Microorganisms are grown in fermenters (large vessels) and used to produce products such as **antibiotics**.

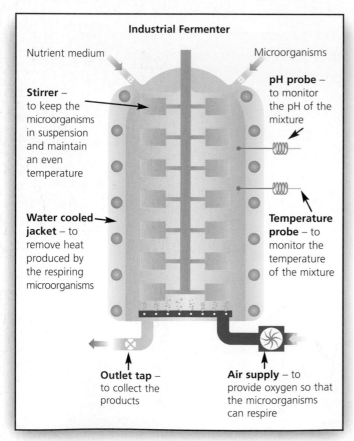

Industrial Fermenter

Nutrient medium

Microorganisms

Stirrer – to keep the microorganisms in suspension and maintain an even temperature

pH probe – to monitor the pH of the mixture

Water cooled jacket – to remove heat produced by the respiring microorganisms

Temperature probe – to monitor the temperature of the mixture

Outlet tap – to collect the products

Air supply – to provide oxygen so that the microorganisms can respire

Products

- **Penicillin** is made by growing penicillium, a mould, in a fermenter. The medium contains sugar and other nutrients which tend to be used for growth before the mould starts to make penicillin.

- **Mycoprotein**, a protein-rich food suitable for vegetarians, is made using Fusarium, a fungus. The fungus is grown on starch in aerobic conditions and the biomass is harvested and purified.

Fuel Production

Fuels can be made from natural products via fermentation. However, all oxygen must be excluded so that anaerobic fermentation can occur.

Biogas, which is mainly methane, can be produced in this way using a wide range of organic or waste material containing carbohydrates.

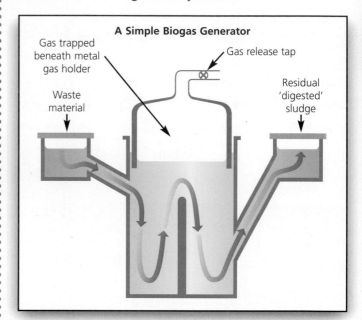

A Simple Biogas Generator

Gas trapped beneath metal gas holder

Gas release tap

Waste material

Residual 'digested' sludge

Many different microorganisms are involved in the digestion of waste material. Waste from sugar factories or sewage works can be used to provide biogas on a large scale. On a small scale, biogas generators can supply the energy needs of individual families or farms.

Anaerobic respiration can also be used to produce ethanol-based fuels from sugar cane juices or glucose derived from maize starch by the action of carbohydrase (an enzyme). The ethanol produced needs to be distilled from the other products of fermentation, and can then be used in motor vehicles.

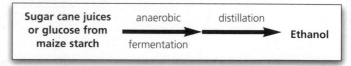

Sugar cane juices or glucose from maize starch	anaerobic fermentation	distillation	**Ethanol**

You need to be able to interpret economic and environmental data relating to the production of fuels by fermentation and their use.

Governments all over the world are looking for ways to avoid a global energy crisis. The combination of dwindling fossil fuel supplies and the long-term damage to the Earth and its climate has increased the urgency to find alternative fuel supplies.

Ideally, the energy sources should be renewable (e.g. the Sun, wind) or use waste biomass from other organisms. Two examples of fuels made from waste biomass are biogas and gasohol.

Fuel	Advantages	Disadvantages
Biogas • Made by fermenting household or farmyard waste in a fermenter or biogas generator. • Biogas is a mixture of methane (up to 80%), carbon dioxide and water.	• Renewable energy supply. • Reduces use of fossil fuels. • Leads to a reduction in by-products from burning fossil fuels. • Cleaner than fossil fuels. • Cheap energy source. • Uses all types of organic waste. • Less damaging to environment than releasing pure methane into the air. • 20 times less damaging to the planet than allowing waste to rot. • Removes unwanted waste such as rotting contents of landfill sites.	• Can take up to one month to produce biogas. • Initial start-up is costly. • Storage facilities are needed.
Gasohol • A biofuel that consists of a mixture of alcohols (such as ethanol or methanol) and petrol. • It is produced from sugar cane, corn starch and fast-growing trees like willow.	• Renewable energy supply. • Cleaner than pure petrol. • Reduces use of fossil fuels. • Less air pollution from products of burning fossil fuels. • More plants, such as sugar cane, are grown to produce fuel alcohol, which reduces CO_2 in atmosphere. • Motor vehicle manufacture receives a boost as cars are converted / new designs are developed. • Cheaper than pure petrol. • Sugar cane industry rewarded financially.	• Consumption up to 25% more than petrol alone. • Burns at higher temperatures than petrol so engines need modifying. • Alcohol can corrode metal so fuel tank needs protecting.

How Science Works

You need to be able to evaluate the advantages and disadvantages of given designs of biogas generators.

Most gas generators are based on two designs – in your exam they may be different shapes but the basic principles will be the same. General features to look out for include...

- an inlet point for waste to be fed into the fermenter
- an outlet for gas collection
- a valve to prevent build up of gas to high pressure
- a stirrer to mix waste and release gases
- a seal to maintain anaerobic conditions (i.e. to prevent oxygen entering)
- a water jacket to keep temperature constant
- an extra tank to collect slurry.

Batch Biogas Generator	Continuous Biogas Generator
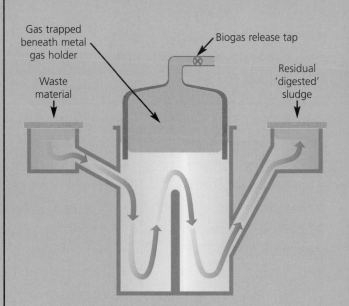	
1 Waste material is put into the tank with anaerobic bacteria. It is left to ferment until all the gas has been produced. 2 The gas is collected. 3 The tank is then cleaned and refilled. 4 The waste or slurry in the tank may be sterilised and used for fertiliser.	1 Waste material is continuously added to the tank to 'top up' fuel production. 2 The gas is collected. 3 This continues for long periods of time without a break in production. 4 The slurry is collected, sterilised and used as fertiliser.
Advantages • Useful for small-scale production. • Can be cleaned easily if contaminated.	**Advantages** • Efficient – don't have to close down fermenter. • Methane produced all the time. • Continuous use of waste materials.
Disadvantages • Takes time and energy to set up. • Can take up to four weeks before methane is produced. • Only small batches of biogas produced.	**Disadvantages** • More expensive to set up as it is more complex. • Needs a continuous feed of waste. • Large amounts of gas need storing. • Can't control temperature so it is slow in winter.

13.7

How can we be sure we are using microorganisms safely?

Microorganisms must only be used if there is a pure culture, containing only one microorganism species. To understand this, you need to know…

- how microorganisms are grown
- why uncontaminated cultures are needed to make useful products
- how uncontaminated cultures are developed.

Preparing a Culture Medium

Microorganisms are grown in a culture medium containing various nutrients that the particular microorganism may need. These may include…

- carbohydrates – as an energy source
- mineral ions
- vitamins
- proteins.

Agar is most commonly used as the growth medium. This is a soft, jelly-like substance which melts easily and re-solidifies at around 50°C. The nutrients mentioned above are added to the agar to provide ideal growing conditions for cultures.

Preparing Uncontaminated Cultures

If the cultures we want to investigate become contaminated by unwanted microorganisms, these 'rogue' microorganisms may produce undesirable substances which can prove harmful.

It is only safe to use microorganisms if we have a pure culture of one particular species of microorganism.

To make useful products, uncontaminated cultures of microorganisms are prepared using the following procedures.

① Sterilisation of Petri Dishes and Culture Medium

Both petri dishes and culture medium are sterilised using an autoclave. This is a pressure cooker which exposes the dishes and the agar to high temperature and high pressure to kill off any unwanted microorganisms.

② Sterilisation of Inoculating Loops

Inoculating loops tend to be made of nichrome wire inserted into a wooden handle. They should be picked up like a pen and the loop and half the wire should be heated to red heat in a bunsen flame, before being left to cool for five seconds. They are then sterile and can be used safely to transfer microorganisms to the culture medium.

N.B. Do not blow on the loop or wave it around to cool it as this will mean it will pick up more microorganisms.

③ Sealing the Petri Dish

After the agar has been poured in and allowed to cool, the petri dish should be sealed with tape (to prevent microorganisms from entering) and clearly labelled on the base. It should be stored upside down so that condensation forms in the lid.

In schools and colleges, cultures should only be incubated at a maximum of 25°C to prevent the growth of pathogens that grow at body temperature (37°C) and are potentially harmful to humans. In industry, higher temperatures can be used to produce more rapid growth.

Example Questions

For Unit 3, you will have to complete one written paper with structured questions.

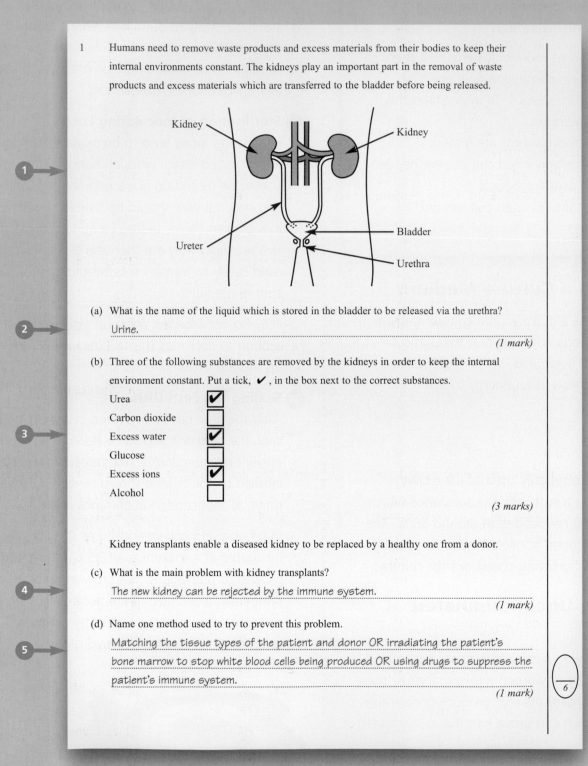

1   Humans need to remove waste products and excess materials from their bodies to keep their internal environments constant. The kidneys play an important part in the removal of waste products and excess materials which are transferred to the bladder before being released.

Kidney   Kidney

Ureter   Bladder

Urethra

(a) What is the name of the liquid which is stored in the bladder to be released via the urethra?

Urine.

(1 mark)

(b) Three of the following substances are removed by the kidneys in order to keep the internal environment constant. Put a tick, ✔, in the box next to the correct substances.

Urea ✔
Carbon dioxide ☐
Excess water ✔
Glucose ☐
Excess ions ✔
Alcohol ☐

(3 marks)

Kidney transplants enable a diseased kidney to be replaced by a healthy one from a donor.

(c) What is the main problem with kidney transplants?

The new kidney can be rejected by the immune system.

(1 mark)

(d) Name one method used to try to prevent this problem.

Matching the tissue types of the patient and donor OR irradiating the patient's bone marrow to stop white blood cells being produced OR using drugs to suppress the patient's immune system.

(1 mark)

6

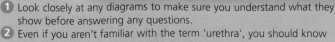

1 Look closely at any diagrams to make sure you understand what they show before answering any questions.

2 Even if you aren't familiar with the term 'urethra', you should know what the function of the bladder is.

3 If you are unsure about the answer to questions like these, eliminate the options that are obviously wrong first.

4 There are lots of different ways of saying the same thing, so the examiner looks for the correct use of key words, e.g. if they marked this question they would be looking for the words 'reject' and 'immune system'.

5 Look back to your answer in part (c) to help you answer this question.

Active transport – the movement of substances against a concentration gradient

Aerobic – with oxygen

Alveoli – air sacs in the lungs. Oxygen diffuses into them and carbon dioxide diffuses out of them

Anaerobic – without oxygen

Biogas – fuel produced from the anaerobic decomposition of organic waste

Blood plasma – the clear fluid part of blood that contains proteins and minerals

Concentration gradient – a change in the concentration of a substance from one region to another

Culture medium – a nutrient system used for the artificial growth of bacteria and other cells

Dialysis – the artificial removal of urea and excess material from the blood (used when kidneys fail)

Diffusion – the mixing of two substances through the natural movement of their particles from a high concentration to a low concentration

Dilate – to widen or enlarge

Fermentation – the conversion of sugar to alcohol and carbon dioxide using yeast

Germinating – when a seed begins to grow

Glycogen – a form of starch in which sugars are stored in the body for energy

Guard cells – pairs of sausage-shaped plant cells which open and close to allow oxygen into the leaf and water and carbon dioxide out (through the stomata)

Haemoglobin – red pigment in the red blood cells which carries oxygen to the organs

Immune system – the body's defence system against infections and diseases (consists of white blood cells and antibodies)

Incubated – grown in a laboratory under controlled conditions

Irradiation – exposure to radiation to kill microorganisms

Lactic acid – a compound produced when cells respire without oxygen (i.e. anaerobically)

Malting – e.g. germinating barley under controlled conditions then drying it in a kiln (oven) – the starch is converted to glucose

Methane – clear gas given off by animal waste; can be used as a fuel

Mycoprotein – a protein-rich food produced from fungi

Osmosis – the movement of water through a selectively or partially permeable membrane into a solution with lower water concentration

Oxygen debt – oxygen deficiency caused by intense exercise

Oxyhaemoglobin – the combination of oxygen and haemoglobin

Partially permeable – a barrier which allows only certain substances through

Penicillin – antibiotic drug used to treat bacterial infection (discovered by Alexander Fleming)

Penicillium – a mould, from which penicillin is developed

Petri dish – a round, shallow dish used to grow bacteria

Root hair cells – found on plants, absorb water from the soil

Specialised – adapted for a particular purpose

Sterilised – free from all microorganisms

Stomata – openings / pores in leaves

Surface area – the external area of a living thing

Tissue type – inherited characteristics present on the surface of tissue cells

Transpiration – evaporation of water from plants (through the stomata)

Urea – waste product of proteins formed in the liver and excreted in urine

Urine – water and waste products filtered by the kidneys

Villi – projections which stick out from the walls of the small intestine. Each villus contains a network of blood capillaries for absorbing soluble food

Wilting – the drooping of a plant caused by excessive water loss through transpiration

How Science Works Key Words

Here are the words that might be used in your exam, with a definition so you know exactly what you are being asked.

Accuracy – how correct or exact something is. The more times you repeat an experiment, the closer the average value (mean) of the results will be to the true value.

Analyse – look at in detail

Apply – relate to, put to practical use

Calculate – work out

Consider – think about

Construct – make, put together

Contrast – look at the differences between

Describe – put into words

Determine – decide, conclude

Discuss – talk about

Evaluate – determine the worth of

Evidence – results of an experiment or facts that you can use to prove or disprove a theory

Explain – put into words

Fair test – a test where conditions are controlled so no factors other than the one you are changing / controlling have an effect on what is being measured

Impact (social, economic, environmental) – an effect

Informed judgements – a balanced view based on information

Interpret – explain the meaning of

Precision – exactness, only a small spread / range in results

Predict – make a good guess at what you expect to happen

Recognise – notice, accept or be aware of

Relate – make a connection to something (like a real life situation or other experiments, etc.)

Reliability – dependability of the results, based on how accurate the measuring instruments are and the methods used to collect them

Sketch – a drawing

Suggest reasons for – think of possible reasons for

Theory – an idea about what will happen

Variables – something that changes during the course of an investigation

Independent variable – the variable you change and have control over

Dependent variable – the variable (output) you measure

Index

Acknowledgements

The authors and publisher would like to thank everyone who contributed images to this book:

IFC ©iStockphoto.com / Andrei Tchernov
p.3 ©iStockphoto.com / Andrei Tchernov
p.5 ©iStockphoto.com / Audrey Roorda
p.6 ©iStockphoto.com / Todd Smith
p.7 ©iStockphoto.com / James Antrim
p.16 ©iStockphoto.com / Gary Caviness
p.16 ©iStockphoto.com / Elena Korenbaum
p.23 ©iStockphoto.com / Rob Gentile
p.23 ©iStockphoto.com / Richard Scherzinger
p.26 ©iStockphoto.com / Leah-Anne Thompson
p.29 ©iStockphoto.com / Ben Phillips
p.31 ©iStockphoto.com / Mikhail Tolstoy
p.41 ©iStockphoto.com / Mike Morley
p.41 ©iStockphoto.com / Alex Bramwell
p.55 ©iStockphoto.com / Ståle Edstrøm
p.59 ©iStockphoto.com / Matthew Scherf
p.68 ©iStockphoto.com / Diane Diederich
p.68 ©iStockphoto.com / Radu Razvan
p.75 ©iStockphoto.com / Andrei Tchernov
p.75 ©iStockphoto.com / Gary Caviness
p.76 ©iStockphoto.com / Nicole Hrustyk
p.86 ©iStockphoto.com
p.89 ©iStockphoto.com / Margaret Smeaton
p.89 ©iStockphoto.com / Bill Grove
p.92 ©iStockphoto.com / Dariusz Sas
p.97 ©iStockphoto.com / Dale Taylor
p.99 ©iStockphoto.com

Artwork supplied by HL Studios.

Text By David Bennahum
Cover & book designed by
Pearce Marchbank, Studio Twenty
Computer production by
Adam Hay, Editorial Design
Cover photos by Ken Sharp

Edited by Chris Charlesworth
Picture research by David Brolan

Copyright © 1993 Omnibus Press
(A Division of Book Sales Limited)
ISBN: 0.7119.3798.2
Order No: OP47534

A catalogue record for this book is
available from the British Library.

Exclusive Distributors:
Book Sales Limited
8/9 Frith Street,
London W1V 5TZ, UK.
Music Sales Corporation
257 Park Avenue South,
New York, NY10010, USA.
Music Sales Pty Limited
120 Rothschild Avenue,
Rosebery, NSW2018, Australia.
To the Music Trade only:
Music Sales Limited
8/9, Frith Street,
London W1V 5TZ, UK.

Every effort has been made to trace the
copyright holders of the photographs in
this book but one or two were
unreachable. We would be grateful if
the photographers concerned would
contact us.

Printed and bound in Singapore by
Singapore National Printers Limited.

Acknowledgements
Thanks to Miriam Bird, for her
excellent research assistance;
Mave Dempster for the videos and
clippings; Mayor Marlene Arpp, Village
of Consort, Alberta, Canada, for her
anecdotes; The National Museum and
Archive of Lesbian and Gay History at
the Lesbian & Gay Community Services
Center in Manhattan; The New York
Public Library for just existing;
Everyone at Music Sales in London:
Chris Charlesworth, Dave Brolan and,
of course, BW.

Photo credits
Bill Davila/Retna: 15, 27
Glen Erler/Warner Bros: 6, 43
Ken Friedman: 26-27
Steve Granitz/Retna: 33, 48
Mick Hutson/Redferns: 10, 24, 32
Chris Kraft/Retna: 29
Bruce Kramer/LFI: 25
Dorothy Low/Retna: 17
Henry McGee/Retna: 40
Marc Marnie/Redferns: 11
Robert Matheu/Retna: 8-9
Paul Natkin/Photo Reserve Inc: 22-23
Chuck Pulin/Star File: 12-13, 19
David Redfern/Redferns: 16, 34
Aubrey Reuben/LFI: 36
Ebet Roberts/Redferns: 14, 18-19, 42
Ken Sharp: 1, 4-5, and covers
Michael J. Spilotko/Star File: 39
Geoff Swaine/LFI: 28
Ron Wolfson/LFI: 30-31, 38, 41, 44-45
Vinnie Zuffante/Star File: 33, 44, 46

About the author
Born in New York in 1968,
David Bennahum graduated in
Literature from Harvard University.
He still lives in New York, surviving
mainly on a steady diet of bagels and
butter. This is his second book (the
first was In Their Own Words: The
Beatles After the Break-up). He is looking
forward to learning how to cook.

k.d.lang

OMNIBUS PRESS
London/New York/Sydney

Few singers have attained as much critical acclaim and demonstrated such talent, yet sold so few records as k.d. lang.

Today k.d. stands on the verge of superstar status and can look forward to a career that will stretch on for decades. Her accomplishments are manifold: in 1992 she won the Grammy for best female vocalist, repeating her 1989 award for her *Crying* duet with Roy Orbison. In the winter of 1988 k.d. brought down the global village; 60,000 people in one arena and over a billion watching around the world, and this 26 year old blew them away. Thousands of athletes poured down from the stands in a spontaneous square dance; k.d. feeding on the energy of 20% of the world's population, building on it, and giving it back. That's a woman who's unstoppable.

Many observers think of k.d. lang as a lesbian, vegetarian, feminist, pantheist, and oh-and-by-the-way, a singer.

She is all of those things, sure; and, yeah, she thought she was the reincarnation of Patsy Cline and, yeah, at times she's said animal rights are more important than human rights. But that is not the source of her power, that's a diversion to discuss over lunch. What fuels k.d. is her divine talent, a voice and presence so profound that, in concert, audiences collapse and beg for mercy.

She has the potential to become the greatest singer since the King himself.

Those who've seen her live know what I'm talking about; the rest, go pick up the video compilation 7 *Years' Harvest*. Watch it twice. By the end you'll be covered with goose bumps, grinning like an idiot, and howling like a wolf.

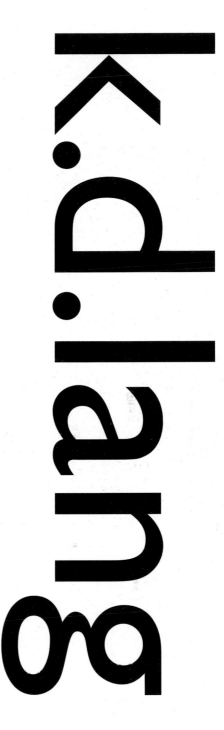

country roads to stardom

k.d. released *Ingénue* in April of 1992. It marked a radical shift in musical style for her, a farewell to the past seven years of hard work as a country singer. Why did she do that? What made her shift from country music to what *Rolling Stone* calls Alternative Lite? Country is more than a music, it is a way of being.

"Country music mixed my reality and my fantasy together because my fantasy was to be a star. I'd always looked on it as a less intelligent form of music, like hip people do, but then I really started to love it. I liked the kitsch aspect but I also liked the value system of it. It's like a mixture of blues and Christian values and it's very extreme. Performance art got a bit boring because there were no barriers to fight against. One of the things I found most interesting about country was the idea of being creative within a structure.

"We've entered a phase where music is very sterile and synthesised and narcissistic, while country music has a real humanistic quality to it. It's dealing with human emotions and I'd say that's something we're all craving right now.

"The time has come for me to let go of the idea of being a country singer... I'm not prepared to make the kind of compromises that would be necessary for me to be accepted by these people. At one time I did very much want to prove to them how much I honestly loved country music, but they made their own assessments whether you're honest or not.

"They loved the fact I brought new viewers in but didn't want to accept or subscribe to my viewpoints – vegetarianism,

lesbianism, things that don't suit the stereotypical role of the female... Looking back it was perfect. I had success, like the Grammy... and yet never had airplay... so you had this huge contradiction, which I thrive on."

Country rejected k.d. because k.d. thought she could "do" country without having to "be" country. k.d would ultimately have to "be" country in order to fit in. And there was no way k.d. would sell herself out like that. Early on she realised that if she wanted, she could go the "pop" route and have success that way but in her view this would be a shallow, transient success. So she chose a music that she thought had roots, a music with tradition, that spoke to a culture. k.d. believed country represented that. And if she wasn't going to change for country, maybe country would change for her. k.d. hoped that she would bring new energy, new juice to a slumbering musical tradition.

Instead, k.d. found herself in a media marketing haze of good ole boys self-righteously proclaiming to be part of a "culture". She discovered that, no matter how hard she tried, these gatekeepers had an agenda that had little or nothing to do with country's roots. But they had power; the power to censor her music, yanking it off the radio waves, no matter how many prizes she won, or with which famous producers she had teamed up. And yank it they did. The toll was high. Of her country career, k.d. never made it above 21 on the *Billboard* singles chart.

She relied on television, record sales and live concerts to get word out and k.d. built up a loyal and devoted base of fans by touring – up to 100 shows in one six month period between July 1989 and February 1990. And those fans were not just loyal, they were devoted.

Country music was originally the white man's blues. It's hillbilly music that speaks to the dispossessed: the suffering of the poor, the abused, the drifters. It's an anthem that poor whites shared from Appalachia to the Ozarks and on out to the deserts of the West. The very name Country & Western implied a folk music rooted in the land and towns which came from the people and was for the people. As with all folk music it was unpretentious and democratic. Country linked communities together in a time before radio or television when live performance, whether professional or amateur, was the only means of hearing songs. Now imagine Dolly Parton and Kenny Rogers. What the hell do they have to do with that? And therein lies sixty years of hypocrisy in a nutshell – from folk to finance.

While wrapping themselves in a patriotic, Christian, sentimental slurry, Nashville (the "capital" of country music) took it on itself to "reform" country and western during the '60s. By dropping the Western and keeping the Country, they cleaned up the hillbilly sound, sucked out the soul and developed a music that sounded more and more like muzak. And sure enough, as rock & roll took over the universe, country languished in a torpor. Ridiculed and parodied as foolish sentimental noise, filled with hackneyed drunks, tough wives, and generic drifters, it no longer appealed. It was a classic short-sighted Faustian pact, selling soul for riches and thus was created a dearth of meaningful country music. And as with any vacuum, it eventually got filled by singers and songwriters who saw a tremendously exciting opportunity. They dreamed of bringing an identity back to country

music. The '80s saw a new breed of country singer emerge, artists labelled "new traditionalists" by the press, and to some measure they succeeded in returning some integrity (and fun) to the music. These singers were determined to respect country's roots while fighting its slow rot and they included Dwight Yoakam, Lyle Lovett, Steve Earle, and, of course, k.d.

Today, country music has surpassed Top 40 as the third most popular radio format in the US, right behind Adult Contemporary and Talk/News. There are now 2,500 country radio stations in the United States. The top album of 1992 was Garth Brooks' *Ropin' The Wind*, beating Guns N' Roses and U2. So it would seem rock'n'roll's met its match. Arguably, the country "ghetto" of the past is gone. You no longer have to leave country to be successful, "crossing over" is no longer a prerequisite to selling more albums. You would think all this has changed the country establishment's attitudes. Perhaps it has, but the pace is glacial, and lesbian-feminist-vegetarian-pantheists need not apply (for now).

k.d. might have waited it out. Perhaps in a few years they would've accepted her. And maybe she wanted to try. But to her surprise that choice was to a large measure taken out of her hands in 1990. It's referred to as the PETA controversy. It ultimately served as the catalyst that drove k.d. out of country.

the west wasn't won on salad!

"My personal thing is animals. Women's rights, of course, but my number one protectionist energy goes toward animals maybe even before women. If people could respect animals, they might respect themselves more."

"We all love animals, but why do we call some 'pets' and some 'dinner'? If you knew how meat was made, you'd probably lose your lunch. I know – I'm from cattle country, and that's why I became a vegetarian. Meat stinks, and not just for animals, but for human health and the environment." *Actual text of the never aired PETA advertisement.*

"The funny thing was that when I made the 'Meat Stinks' statement, they said they banned my records. Well, the truth was that no one was playing my records anyway."

"Warner (Records) thrived on the publicity. At first they thought it might hurt sales. But those men in the trucks out there, those aren't the guys you see at my concerts."

"The vortex of the controversy was in my hometown. It wasn't so much the criticism of the press. It was the personal attacks on my family that was really painful... I went from being Canada's little queen to all of a sudden having the whole country shift against me. It was a little scary."

"It's basic human nature to look for something to hate in somebody... They're just waiting for you to do something wrong."

k.d. grew up in cattle country, Consort, Alberta (population 650 to 714 depending on who you ask). She grew up with meat on the table, meat in the fields, and meat in their shoes and jackets. As long as she lived at home, she ate the meat on the table. She was always a vegetarian inside, but, as k.d. puts it, "it would have been a little tough to grow up where I did and say 'mom, don't cook animals'."

k.d. won't hesitate to tell you not to eat meat. She cares enormously about this issue. But she isn't militant in the sense of forcing those around her to change (although she claims she'd have a hard time taking a lover who wasn't a vegetarian, or trying to be). Whenever she tours, her band and crew are fed, and this always includes meat. She wouldn't ban meat from her own circle.

Picture this. On her own, not thinking it's a big deal, k.d. tapes a public service announcement for a group she believes in: People for the Ethical Treatment of Animals (PETA). It's a simple spot: k.d. standing by an adorable mammal of the bovine variety, smiling and telling the camera that "Meat Stinks". k.d. thinks that it stinks for a lot of reasons, not the least of which is the fact that raising beef destroys thousands upon thousands of acres of topsoil, uses tons of pesticides and fertiliser, and generally makes no environmental sense. All this stands aside from the spiritual question of what it means to kill an animal.

k.d. taped this announcement without telling Sire Records. After all, why bother, plenty of people support environmental causes. It has nothing to do with music. It's altruism: a celebrity giving her name to a cause she believes in. It happens all the time. And in this case the logic is sound, it's not like she's advocating the legalisation of drugs or the slaying of first born. Who is going to complain? Well, with hindsight, impeccable as it always is, it makes perfect sense that an uproar would blossom over this "controversy". After all, isn't country music beef eater's music? No? Because if you think it isn't, then there is something afoot here. Not eating meat, how dare she! Why, the foundations of the nation are at stake. Or more bluntly, as the highly esteemed North Dakota Beef Commission, that arbiter of cultural mores, put it (on a billboard) "THE WEST WASN'T WON ON SALAD!" Of course it wasn't. It was won by 1) men 2) carnivores 3) Americans (not Canadians) 4) cows? Forget about the genocide of the Native Americans, that's irrelevant, the point is, the West was won. No one should forget that the very foundations of the Continental United States rest upon the rock-solid-God-fearing bellies of Beef Eaters. And no lesbian-feminist-vegetarian-pantheist-Canadian (singer) can tell them otherwise. To do so is treason.

If you're beginning to think that all this is a little exaggerated, then you don't quite understand the meaning of "waiting for an excuse to do her in". Because at its root that's what this is all about. The PETA ad was never aired because this "stink" arose before it even made it on TV. (That highly esteemed investigative news show *Entertainment Tonight* was the first out with the "news".) From that point on this surrealist tragi-comedy spiralled out of control into the realm of the absurd.

Apparently, the logic went something like this: k.d.lang wants to destroy the livelihood of our listeners, therefore we will destroy her livelihood first in a massive pre-emptive strike, banning her records from the air waves. Well, seeing that the ad was never aired, it's hard to see how k.d. actually posed such a threat. But this country singer apparently held the fate of meat eating America between her fingers.

Memos made their way around stations. KRVN-radio (in Lexington, Nebraska, serious cattle country by the way) announced in a memo: "Under no circumstances will anyone on this staff be allowed to play music by k.d.lang until (she)... renounces her *fanatic* anti-meat philosophy". Too bad KRVN only played her music two or three times a week to begin with. As to the listeners, and what they thought, the reaction was decidedly dispassionate. A poll taken by another radio chain (Great Empire) found 30% wanted her off the air, the rest didn't care. Now, not all radio stations cowered in fear. Some prefaced all k.d.lang material with the sounds of cows mooing, as a sort of thumb-in-the-nose to the stations banning her.

If there is any happy ending to this controversy, it's that, in the immediate period after the ban went into effect, k.d. saw her record sales quadruple. She pulled in legions of students and alternative listeners that otherwise would've assumed k.d. was just another country singer. She actually broadened

the base of her audience, branching into new demographics, to use a marketing term.

But, in a deeper sense, there is no "happy" ending. The sign welcoming visitors to her hometown used to say "HOME OF k.d.lang," after the boycott it read, defaced, "EAT BEEF DYKE". Her mother, who still lives there, received threatening phone calls and letters. k.d. asked her to move, but she refused, taking it all in her stride. This attack on her family hurt her more than anything else. The rest was commercial, political, but this was personal. Her sister, Keltie, described it: "It was like you'd imagine a TV movie about something hateful in a small town in the South to be". Now k.d. can look back on the experience and try to rationalise why her hometown turned on her: "It's a very small town mentality, to criticise and hate when someone's done well... in Canada they eat their young. And I'm told that in Britain they do the same thing. America doesn't suffer from that, for some reason."

The deeper reason for the boycott had nothing to do with whether or not beef jobs were at stake. It's that country is not simply a genre, it represents a philosophy, and to some degree a politics as well. What this experience did was highlight how impossible it was for k.d. to sing country without "being" country. One, for now, can't separate the two. And that, more than anything, drove home the wedge of seeming incompatibility between k.d. and the country "establishment." They weren't going to let her in. And they'd seized on a never-aired advertisement as the rallying cry to drive that point home.

k.d. could sing all she wanted, work with the finest of the country greats, but there was no way that she would ever be accepted. And there was no way that k.d. was going to change. So she walked off the mat, taking with her an impressive legacy.

This decision was a major one. Her country recordings are sublime, her live renditions transcendent, and her commitment as an artist truly sincere.

That such a talent could be driven out is a loss for country, but certainly only a beginning for k.d. What k.d. left behind are four country records *A Truly Western Experience, Angel With a Lariat, Shadowland,* and *Absolute Torch and Twang.* Each album was a quantum leap above the next. Her pace of change was so fantastic both in quality and magnitude that one could only imagine what would come next.

k.d.'s challenge to country was not simply musical, it was her interpretation of the music, and underneath that, the lodestone of her sexuality, which simmered below the surface, hiding behind a comic androgyny that allowed her to blow wide open the limits of country. She blasted those doors down, and the howling wind scared the shit out of the establishment because, and this is the killer, *the people loved it.*

the big skies of alberta

"I had really supportive parents. I remember my mother saying 'You're very handsome.' I really loved that she said 'handsome', because, you know, again that's breaking stereotypes. You're very handsome, and you don't need to wear makeup. You have beautiful skin."

"Whenever an intelligent, strong woman came to town who looked independent... it was like, 'I want to get to know you'."

"My first hero was Maria in the *Sound of Music*. I loved her wholesomeness and her happy-go-lucky attitude."

"In high school I took an aptitude test that said I was 98 percent guaranteed to be a mechanic."

"My father treated me like a tomboy... He bought me a motorcycle when I was nine; I've been riding cycles for 22 years. I was a marksman; I used to shoot guns with him... I remember him getting me an electric guitar for Christmas when I was in grade six."

"I was really close to (my father) before he left. Extremely close actually. And I ran into him a couple of times after about eight years... I just thought, well, I'll carry on... A wound is just a highway to a new and enlightened confidence, basically... Everything you go through is marking your soul."

Within months of Kathryn Dawn's birth on the second of November 1961, Fred Lang moved his family to Consort, Alberta, and opened Lang's Pharmacy. His wife, Audrey, taught second-grade. The family split when Fred walked out 12 years later, leaving Audrey to raise four children: Jo Ann (now a married nurse with four children), Keltie (studying fine arts in Calgary), and John (who "does" stained glass in Vancouver now, according to mom). They became Kathy's first audience. Hyperactive, and born to perform, Kathy worked the local scene, singing at school functions and weddings at five. Everyone knew what she wanted to do: sing. So Audrey got her braces, telling Kathy that if she was going to be on stage, she'd need straight teeth.

Kathy used to tear through the town on her motorcycle, pretending she was a cop. Thankfully, around age 10, Consort paved its first road. There were no movie theatres, one TV station, one bar, one drugstore (guess whose), and no police. Consort faced the open prairie, surrounded by fields and sky. People relied on their imagination for entertainment. Kathy grabbed any new morsel of information she

could get her hands on, stoking her mind, "An album cover would be like a movie – a whole other dimension I would travel in, like stepping through the looking glass. Everything I ever did was part of the development of my imagination and lust for discovering new cultures and new sounds." She took what she got as a kid, and it made her very open to art. k.d. claims to embrace what is kitsch and mediocre, unafraid to embrace the "geek inside" herself.

Kathy's first ambition was to be a roller derby queen. The only interesting thing on TV, she decided her fate would play itself out in the high stakes world of the roller derby. That phase passed. She went on to sports, playing everything, succeeding even in joining the men's basketball team. That phase passed. Behind it all lurked music, drifting in and out of her mind. The family studied classical music. Every week mom drove John to a convent sixty miles away. While he practised his scales on the piano, seven year old Kathy sat in the lobby waiting, listening. She learned the scales. Soon Sister Xavier was teaching Kathy how to sing. Kathy quit when the Sisters refused to let her play fusions of pop, jazz, and whatever else floated in (around age 10). Years later the nuns would return to see k.d. in concert.

Kathy's fatalism bordered on hubris. At one point she wrote out a set of lyrics in purple magic marker and sent them to Anne Murray, at that time Canada's leading female country singer (k.d. eventually broke Murray's eight-year lock on the Juno awards), telling her that "You have my permission to write music for these lyrics." Ms Murray never replied. When Keltie went off to college, Kathy sent her a pack of *Charlie's Angels* bubble-gum cards and signed them, telling her to keep them, since the signature would be worth something one day. In k.d.'s own words, "I had a view of what I wanted to do when I was in the womb. It was never a question that I'd go into music but the only thing was what kind?" Wombs with a view are rare, but in Kathy's case, clearly fortuitous.

Evidently, Fred's abrupt abandonment of the family marked Kathy more than any other single event of her childhood. She's reluctant to talk about it, keeping family matters private. It hurt her mother terribly, and this more than anything upset Kathy. To this day k.d.'s never been reconciled with her father. Once, at a concert in Edmonton, someone saw Fred in the audience. He showed up to watch k.d. sing, tears running down his face. The wounds remain. According to k.d., Fred shaped her vision of love as an ecstatic yet unreliable experience. The agony and ecstasy of love became a dominant theme in k.d.'s work.

a cow punk gets into the pen

"When I first got to Nashville, I was given a pink handbook on how to be a country and western star. Section 1A, the first rule of country and western stardom, is, 'The higher the hair, the closer to God'."

"(Androgyny is) synonymous with k.d.lang... a polite way of having people speculate."

"I think country radio suspected she was a lesbian, and even if they weren't sure she was a lesbian, the image was all wrong. They weren't about to put k.d. on a pedestal and use her as a role model."
Larry Wanagas, k.d.'s manager.

"Ultimately my artistry dominates anything I do, and that encompasses the transcending of any sort of gender categorisation in my music. 'Middle of the road' is what fascinates me...(it) is the safest and the unsafest to be, because you're that much farther away from the ditch but you're that much closer to being hit by oncoming traffic! What intrigues me is being alternative and being completely conformist at the same time. And that's actually how I try to live my life."

"I knew I was going to get trapped, with people expecting me to be strange and not paying attention to the voice. Every time I'm on TV they want zany and crazy k.d."

In 1979, the year Blondie and New Wave filled the airwaves, k.d. came to Red Deer College (in Red Deer, Canada). She'd left home to study music in a two year program. After eighteen months, k.d. ditched school to pursue music full-time. A jock (she was placed eighth in Canada's javelin competition), filled with artistic ambitions, k.d. hungered for intellectual stimulation; she started exploring various media with intensity. What she got into was definitely not what Red Deer College thought music should be.

k.d. hung out with punks and discovered performance art, and soon her interests diverged from those of the school. At one point she spent twelve hours re-enacting Canada's first open heart surgery operation. She sat out there on stage, using vegetables as organs, and meticulously went through all the steps while an ever-changing audience walked in and out. She'd go to parties and sing wildly while accompanying herself on the guitar, freaking some out, and attracting others. Or else spend time in bags, building strange sculptures. She checked it all out, sampling what was out there, looking for her muse. She took casual work, including a stint as a truck driver.

k.d. always knew she'd be famous, the question for her was which path to take. Performance art and punk attracted her, mainly because she loved the energy of these genres. But, as Sid Vicious discovered the hard way, music without boundaries can consume you, and k.d. saw that. She wanted something else, but wasn't sure what it was. Back then she didn't much like country, she hadn't even listened to it at home. Her family raised her on Broadway musicals, classical, Janis Joplin, and The Allman Brothers. She thought she might be a jazz singer, toyed with that for a while. But jazz is not exactly the path to fame and stardom. Then she discovered Patsy Cline.

Patsy Cline is arguably country music's most enduring heroine. She died aged 31 in a plane crash in 1963, the same year Kennedy was shot, and to many of her fans Cline's death is the *real* marker of that year. Her ethereal voice and emotive songwriting testified to country's true potential at a time when gimmick-ridden singers were on the rise. Years before her time, Patsy represented a new and honourable "Nashville Sound", an attempt to take on rock'n'roll by allying its free spirit with something deeper than what many saw as the trivial preoccupations of pop. But Patsy was no sell out; she was sophisticated and glamorous, and she was by no means a parody. She became one of the first country singers to cross over into the pop charts, and hits like *Crazy* and *I Fall to Pieces* have become lasting standards.

Doug Newell, a teacher at Red Deer's Drama Department, saw Lang play guitar at a party and offered her a part in a country musical he was staging entitled *Country Chorale*. Newell wanted a singer who could play a character loosely modelled after Patsy Cline. At the time lang didn't even know who Patsy Cline was, but that didn't stop her. She threw herself into the role and was transformed and in the process she discovered another style of country music. When the show ended she announced that she'd no longer be Kathy Dawn. From then on she called herself k.d. because k.d. is "a name, not a sexuality".

k.d. dreamed of fiery plane crashes and took it as an omen. Not simply inspired by Patsy Cline, she became the reincarnation of Patsy Cline, or so she said (now this phase is more delicately referred to as the "creative" reincarnation of Patsy). The reincarnation of Cline later took it upon herself to name her band The Reclines. Whatever one makes of this, k.d. dived into the country scene with a vengeance. She

studied Cline's records and transformed the genre. She'd found her muse. k.d. decided that country "paralleled the environment I grew up in… bake sales, the rodeo, the ordinary guy and girl. Hating the music at first, and then coming back to it again, realising how cool it is, gave me a kind of distance." With that in mind, k.d. moved towards her self-proclaimed destiny, the one she knew as a child, to captivate the world by song.

Drawn now irrevocably towards country music, Lang answered a newspaper ad for a 'singer in a swing band', showed up for an audition at Bumstead Recorders Ltd. in Edmonton, Canada, and blew the room away. The studio's owner, Larry Wanagas, was looking for someone who could give country "a good kick in the ass" and front a band he managed called Dance Party. The band didn't last but Wanagas signed her on, and he remains her manager to this day. "Anyone who can sing that well lying on the studio floor… should be incredible standing up," was one of his first observations. That was April 1983.

Wanagas suggested that k.d. form a band and the first of several editions of The Reclines was hastily assembled. Over the next two years they performed regularly and changed personnel several times. Keyboard player Stewart MacDougall wrote some songs of his own and helped k.d. as best he could before giving way to Ted Borowiecki and then Mike Greber. Two bass players, Farley Scott and Dennis Marcenko, came and went

before John Dymond made the position his own, and Miche Pouliot took over from original drummer Dave Bjarnson. Guitarist Gordon Matthews stayed the course during the comings and goings, but slide guitarist Gary Koliger and organist Jamie Kidd were dropped after becoming surplus to requirements. In 1984 k.d. was joined by violinist Ben Mink who would go on to become her principal collaborator.

Curiously, although Mink is generally featured on either violin or mandolin in The Reclines, his background was in hard rock. He began by doing sessions in Canada in the late Seventies, and became a member of a Canadian AOR band called FM. In 1982 he was invited by Canada's biggest rock outfit, Rush, to contribute to their *Signals* album but a later stab at establishing himself as a solo artist went awry.

k.d. cut 'Damned Old Dog' coupled with a Cline tribute, 'Friday Dance Promenade', for Bumstead in '83 but only 500 copies were pressed. Today a mint condition of this single can fetch over £30. By the end of that year she sang on *Sun Country*, a country TV show beamed around Canada. You can catch the clip of k.d.'s TV début: she appears wearing a black dress, studded with three rhinestones (a reference to Patsy Cline) and hair that's only mildly butchered. But that was already too much for some. There was concern that the audience wouldn't like her dressed that way. Compared to her usual performances, k.d. looks positively subdued here. But that wouldn't last long. In the time between releasing 'Promenade' and cutting *A Truly Western Experience* (for $5,000), k.d. created the country persona that later, in '88, she'd try to overcome with the release of *Shadowland*. k.d. came of age in Canada, and that's a long way from Nashville, Tennessee. One day, she

knew, she'd have to face the boys down there. Problem was that by then, in the relatively less doctrinaire world of Canadian country, her image was cast, at the expense, she feared, of her voice and potential.

What emerged in '84 was a singer who mixed country with performance art, punk, and new wave. Who would ever have thought to mix that with country? Somehow it worked. k.d. refined this act over dozens of concerts. She toured all over Canada with the ever-changing Reclines, building a fan base and getting word out. When she released *A Truly Western Experience*, her fame already stood in disproportion to album sales. That's because k.d. came into her own on stage. Often shy and withdrawn, wary of the press, k.d. surprised everyone when she walked out and faced the audience. She has the ability to feed off the energy of her audience; she's as strong as the people watching her. With her voice and fiercely entertaining act, there was little to hold the audience back. What did perhaps get in the way was her "in-your-face" attitude. Audiences often felt uncomfortable with that kind of frenetic fun at first; they found k.d. intimidating or, as they put it, "weird". That's a common reaction to the early k.d. but the weirdness overwhelms the music.

"Weird" is simply a way of collapsing what k.d. attempted to do on stage, but it marginalises and dismisses the source of her creative interpretation: she refused to accept the stereotypical role of women in country. Take a look at *Poly Ann*, a track from *A Truly Western Experience*. The video alternates between shots of k.d. in a pink-white-and-brown spotted-polyester suit-dress from the early Sixties (her

"Neapolitan dress", as she calls it) parodying a housewife, and clips of her hungry band banging forks on a Formica table. Where's dinner, woman? is the visual text. k.d., in full flip wig (*sans* glasses), segues out of Nena Hagenesque squawks, Beach Boys' medleys (replete with surfing band members), and Rockabilly riffs. The tune is a paean to the virtues of processed home "cooking"... so much for staying up late and cooking for your hungry man. She tosses breadsticks in the air and presses a slice of Wonder Bread against her heart. It all ends with k.d. (now back in glasses without lenses) ripping off her wig and revealing her shaved head, anathema to old style country and a spit in the eye for the 'higher the hair, closer to God' brigade.

When k.d. first appeared on *Sun Country* she looked woefully uncomfortable and restrained in her black dress, but the hysterics of '84 leap out when she does *Pine and Stew* a year later, back on the same show. Apparently she was so nervous and depressed that, days before the show, k.d. chopped her hair to bits. What peers out across the monitor is a defiant k.d., freaked out by what she calls her "celebritism", looking like Pee Wee Herman in Buddy Holly glasses. This is a totally "straight" country ballad that's completely reversed by k.d.'s stretched out lyric, "mentaaaal...y anguished." She mimes it out, tapping her fingers on her bony head, looking quite "mental", and quite devastatingly exciting. Talk about getting the good 'ole boys off their arses, this is the work of an explosive artist. Next k.d. received Canada's highest award, the Juno, for Most Promising Female Vocalist. She got there based on her first album, *A Truly Western Experience*., and a single, *Friday Dance Promenade*. A ham to the end, and a savvy media operator, k.d. showed up to collect the award in a wedding dress, pledging to be true to her art.

Slowly, inevitably, like a moth drawn to the light, k.d. circled closer and

closer to the big time, until her touring led her to New York City and the Bottom Line Club. In the audience was Seymour Stein, boss of Sire Records.

"I was just transfixed," he said later. "She was wearing country square-dance clothes and real short hair, but you could close your eyes and imagine her singing anything – show tunes, R&B, hits from the 50s, country classics."

"I was playing at the Bottom Line and Seymour Stein came backstage and asked, 'Do you know "Ballad of a Teenage Queen" by Johnny Cash and "She's no Angel" by Kitty Wells?'" reported k.d. "I just looked at him because... Madonna, Talking Heads and The Pretenders – he's signed all of 'em... I just looked at him. It was love at first sight!"

The Bottom Line is a New York institution, and when k.d. showed up there in '85, she knew this gig was a milestone for her career. The audience, littered with industry execs, Seymour Stein being but one, Mary Martin, then director of A&R for RCA records was another, came to see this Canadian "cow-punk". She'd scored a three-quarter page write-up in *Rolling Stone*, announcing the concert. So the buzz was out, and anticipation high. The higher the mountain, the bigger the stakes, the better k.d. gets. Needless to say, by the end of the set, they all wanted k.d. In the end she went with Stein and Sire Records. To this day, k.d. has no regrets about her choice. She

claims to love her record company, and acknowledges them as artists in their own right.

What Stein saw was k.d.'s potential. He wanted to unclutter k.d., removing the veils that kept the audience at bay: the shaved head, lensless Buddy Holly glasses, and dorky outfits were problematic. To expand her audience, k.d. abruptly segued from one image to another, ditching her more overt performance art roots; she shifted into high gear androgyny. She'd been overtly trashing the female stereotype by dressing hideously, and parodying male expectations of what a female country singer should look like. Now she got subtle. k.d. discovered an uncanny ability to look like (and sing like) Elvis: slick, sharp, and unfeminine without being negative; a sensually masculine androgyny that would give k.d. a whole new range of freedom onstage.

Sire Records wanted to bring k.d. to America with a splash. *A Truly Western Experience* wouldn't do it. Time came for another album, her "major label début". For better or worse, they decided to send k.d. to London to work with Dave Edmunds, erstwhile producer of The Stray Cats, former Rockpile guitarist and vocalist, and all round *éminence grise* of rockabilly. The match had as much to do with Sire's relation

ship to k.d. as with their desire to work with Dave Edmunds. Whether or not artistically this made sense came to light all too soon... just about the moment k.d. walked into the studio and started working with Dave. The result, christened *Angel With A Lariat*, emerged from the studios a month later, at the end of June, 1986.

"It was so stressful," k.d. said later. "Part of it was that (Edmunds) didn't understand what I was doing and I was just so hyper and enthusiastic and overly emotional. I fought everything he said, whether it was right or wrong. I just wanted to get my record out and I wanted to be a big star right away.

I like that record now, but I hated it for years. Tons and tons of reverb, 150 milliseconds on everything."

k.d. stormed America, culminating with the Sire contract, and figured phase two (the conquest of Nashville and America) would begin with the release of *Angel With A Lariat*. That didn't happen. *Lariat* sold 460,000 copies worldwide in 1987, in spite of the fact that there was no hit single and radio shunned the album. And without airplay, k.d. failed to reach a large audience. So k.d. and her management team developed an alternative approach: getting on TV, touring, and using award ceremonies as a vehicle.

Lariat is not a bad record. It's not a great record either. It's the victim of the overly high expectations that everyone, including k.d., had for it. Accusations flew around: people claiming Dave Edmunds was at fault. But truthfully, the record was a flawed collaboration on material that only held up well when performed live. k.d. can take anything and turn it to gold in front of an audience. On tape the record sounded loud, as if a wall of noise separated the listener from k.d.'s voice. Why suppress that asset? The rockabilly theme layered over the range of k.d.'s voice and forced it into a narrow, frenetic, band. But in concert that wall dropped off, and the natural quality of her voice and stage persona took over. For the millions of potential

fans who never saw her perform, *Lariat* remained a rather unpalatable choice to pull out of the bin.

With its release in 1987, k.d. toured heavily. The tracks that did best were *Turn Me Round* and the Cline tune *Three Cigarettes In An Ashtray*. In *Turn Me Round* the rockabilly influence drives the tune at warp 10. In this case it works to get the audience moving, whether or not you'd want it blaring out of your stereo is another matter. k.d. played *Turn Me Round* at the Winter Olympic's closing ceremony in 1988, drawing thousands up off their feet. But, undeniably, the standout was k.d.'s rendition of *Three Cigarettes In An Ashtray*.

The way k.d. worked that song became legend. *Cigarettes* is a classic country tune, and normally it's played "straight", meaning you take the story of a broken heart seriously, serenely, with great reverence. k.d., because she can sound just like Patsy Cline, has the freedom to bend this standard, and bring to it something it deserves: some comedy. A problem with country in general is that there is a temptation to be very serious about the genre and accept the fact that women usually get screwed over, their hearts mashed under the heel of a snakeskin boot. *Cigarettes*, the way k.d. does it, brings power back to the woman. She'd do it with cut off cowboy boots, that symbol of Western masculinity (all you Freudian symbolists out there, go figure), the usual short (but not bald) hair, and no glasses (Sire canned that accessory).

Sitting at a table, with an ashtray, she'd sing to the two cigarettes burning. Supposedly taking place in a bar, the two lovers watch their cigarettes burn, idyllic. But alas, a comely lass ventures forth and places her cigarette in the ashtray – now there are three (at this point k.d. drops in a third cigarette). Her lover leaves with the lass, and the woman stays behind, supposedly heartbroken (k.d. takes out two cigarettes, leaving her own behind). Okay, it may sound banal, but in the context of country music, it electrified. Audiences loved it. She could belt it out sitting down, and, for those who'd missed her singing lying down, that was rather impressive.

k.d.'s performances got noticed. *Angel With A Lariat* racked up some awards, the most noted being another Canadian Juno, this time for Best Country Female Vocalist. But Canada was history at this point, the purpose of *Lariat* was to take over America, Nashville in particular. 1987, apart from *Ingénue*'s release, stood as the most delicate time in k.d.'s career. She'd come on strong, Sire had signed her, she'd released another album. It failed to do what was intended. So how to pick up the pieces? Where could k.d. go from here? Knowing k.d., she didn't take the easy way out. No, she ran headlong into the belly of the beast... Nashville.

welcome to music city, u.s.a.

"You look like a boy, dress like a girl, and sing like a bird."
Roy Acuff, announcer, Grand Ole Opry.

"We admire k.d. for her voice and because there is nothing phoney about her... I'm tied to this costume, this persona."
Minnie Pearl.

As part of her promotional blitz for *Lariat*, k.d. booked a spot on the *Tonight Show* with Johnny Carson. Carson, the master of soothing late night hypnosis, used to lull around 13 million Americans to bed every night. Getting a spot on his show was *de rigeuer* for any artist seeking to introduce herself to the great slumbering American mass. That night, along with k.d., Johnny invited Minnie Pearl, one of the grande-dames of country. Minnie represented the trials and tribulations of old-school country: hyper-feminine, made up, hair quite close to God. She found in k.d. a liberation from the binding accoutrements of country and the freedom "we all wished we had". It's not surprising that the two hit it off that night, two generations of country women coming together. Minnie offered to get k.d. a spot on the Grand Ole Opry.

The Opry stands as the holy of holies, the temple within which resides the spirit of country. The first radio broadcast from the Opry was in 1924. Today Opry broadcasts reach millions and the venue itself seats 4,400. It stays faithful to its small-town roots. Every broadcast spans the nation, but the announcers still read homey ads for local sponsors in Nashville, just like they did in the 20's. After moving the Opry and expanding it, the owners inset into the new stage floor the

original Opry stage. Performers today stand on the very same floorboards that supported Patsy Cline in the 50's and 60's.

At one time, Nashville was known as the Athens of the South. Today its claim to fame rests on its singular role as caretaker of the relics. Here you can visit the Country Music Hall of Fame and see such marvels as Elvis' gold Cadillac.

This city of 500,000 cannot live off country alone, but, judging from the sights, it appears they'd like you to think otherwise. All this could be perceived as endearing, quirky, uniquely American. When it becomes big-business though, people take it very, very, seriously.

k.d. faced a unique anomaly in the music world. Nashville ("Music City, U.S.A."), and its community of record labels, producers, agents, managers, live in a world that is completely self sustaining. The Nashville labels are subsidiaries of larger corporate entities in Los Angeles and New York, but in a remarkable financial arrangement, Nashville subsidiaries keep all profits themselves, sharing none with the head offices. In fact, these "head offices" are so only in name. At a time when Nashville artists sell more albums than rock'n'roll artists, the river of money coursing through town reaches flood stage.

The stakes in Nashville are high, and if you screw up, you can't hide behind the fog of corporate management. Millions may be pouring in, but Nashville still runs out of Music Row, a collection of wooden one storey homes all on the same quiet street. In these "offices", business goes on as it has for decades, by word of mouth, handshakes, a clubbish atmosphere dominated by white Christian men. These men check you out, not just musically, but with the perennial question in mind, "will you fit into the club?" Now that so much dough is rolling in, the pressure to take on "corporate management techniques" keeps increasing with every million.

These men face a sea of change built upon on the fruits of their own success, and they're scared.

k.d. walked into this world, and faced Nashville as a guest at the Grand Ole Opry in October 1987. Minnie Pearl pulled through for k.d., and that day, in front of a capacity crowd, k.d. presented her calling card. The audience howled for more, demanding an encore, and k.d. gave them what they wanted. She sang *Crying* and brought the house down.

Earlier that year, Roy Orbison and k.d. got together to record a duet of *Crying* because he loved her voice. k.d. didn't want to do it at first. She didn't get it, and k.d. gave Orbison a choice: either him or her singing, but not both. Thankfully this dementia passed, and k.d. realised "it's Roy Orbison that you'll be singing with, you goon". It led to her first Grammy award, in 1988, for Best Vocal Collaboration. The record is chilling, their voices together creating a unique blend of emotion and pathos that affected k.d. deeply. In Orbison she acquired an artistic mentor and spiritual friend. After his death, she sang *Crying* at the Grammy Awards, and, well, just listen to her. It's incredible.

Still, Nashville wasn't thrilled with k.d. They sniffed a rat, suspecting a false country singer lurked behind that alto voice. With every breakthrough, every award, Nashville's resistance rose proportionately. So k.d. decided to take them on, on their own territory. She'd give them a record that would surpass, a record so pure, so clear, so perfect, that no one could deny her. She would do country better than country could. She would do *Shadowland*.

honky tonk angels

"After working with k.d. for a while, I didn't need to take my pills. She was medicine, invigorating therapy." *Owen Bradley, producer of* Shadowland.

"I think people were suspicious (of k.d.). There's still a lot of good ole boys in Nashville and if they think they are being wanked they aren't going to like it. I thought I'd get her and Owen Bradley together and get them less suspicious." *Mary Martin,* RCA Records.

"(Working) with Owen was trying to finish the whole Patsy Cline thing. I was real serious about Patsy Cline and was indebted to her for inspiring me to do the whole country thing."

***"Shadowland* came out at an odd time in my career...It wasn't strategically planned. I simply had the opportunity to work with one of my idols, and took it. That's what's so special about the record. It was a gift."**

It's unclear who pursued who in the making of *Shadowland*. The official version has it that Owen Bradley, long retired, sitting in his hospital bed, saw k.d. on *The Tonight Show* and, struck by lightning, decided to come out of retirement to produce her next album. Other versions put it that Bradley was approached a number of times before he agreed to produce *Shadowland*. Bradley stands as a giant in the history of country music. He's one of the few people in Nashville that no one snickers and gossips about. He gained notoriety by producing countless records, all contributing to that "Nashville Sound". He produced Patsy Cline's records, along with Minnie Pearl's. At first glance, who better to work with? He sculpted k.d.'s inspiration, Patsy; no one questions his country credentials. And what a coup for this 27 year old to bring out the venerable Bradley from retirement. But then... isn't he an architect of the "Nashville Sound?" Didn't k.d. define herself as standing in opposition to the standard assumptions of country?

Shadowland reveals this edgy conflict. k.d. excused The Reclines, taking on Owen's studio musicians (the "Nashville String Machine"), and proceeded to cut an album aeons away from *Angel With A Lariat*. As the grand finale, she got Kitty Wells, Loretta Lynn, and Brenda Lee to sing with her on the final track, *Honky Tonk Angels' Medley*. These grande dames of country sang together for the first time, and for

a newcomer. Could there be any greater testament to k.d.'s ability to be country? The reviews sniffed out her attempt to take on this smoky, sultry world of country "weepers" as a mixed blessing. Technically marvellous, one critic stated that if Dolly Parton had recorded this album, Nashville would stage a ticker-tape parade. But k.d. is no Dolly. All her energy, the gifts she brought to country, dropped out, leaving behind that voice. And yes, that voice carried the day, but her spirit, her interpretation of country, where did that go? Was this no more than a filibuster in the unending debate of whether k.d. was country enough for country?

Shadowland found little airplay. Not because people didn't like it. The album sold over a million copies. No, this had to do with the country stations and their programmers. Like with Lariat, they chose not to play Shadowland, even though it carried marvellously. The production, smooth and clear, featured k.d.'s voice using the instruments to support it, not overwhelm it. A trucker driving through the night couldn't go wrong listening to this crooner serenade him. Too bad the DJs didn't see it that way. They figured k.d., still too "controversial", wouldn't be liked by the country public. That's a huge put-down to the public, a condescending self-serving denial of their intelligence.

The two tracks that stood out from the compilation, Busy Being Blue, and Honky Tonk Angels' Medley, attracted attention for totally different reasons, and this perfectly reflected k.d.'s dilemma. Medley just stared out at you, a remarkable coup, not a musical coup, just proof of how other Nashville artists respected her voice and the lengths to which they'd go to help k.d. succeed. Blue isn't a country standard, and the way k.d. did it wasn't trying to be more country than country. Not

surprisingly, it's the best track on the album: a combination of jazz-blues and country, it has emotion and groove. It expands the potential of country while honouring its roots. It demonstrated k.d.'s gift, but country still wouldn't accept it.

The disappointing reception to Shadowland finally put to rest any hope of k.d. trying to fit into country by toning down her spirit. From then on, k.d. moved back to "progressive country". They could ignore her tracks on the air, question her integrity and intentions, and put her down, but she'd go on with her work. The result, Absolute Torch and Twang stands out as k.d.'s finest country album, in some ways even finer than Ingénue. Adversity led to inspiration. k.d. always said she thrived off opposition.

"I chose the words 'torch and twang' (because) I'd love to marry ballad jazz and country. Those are the types of music I'm most passionate about."

k.d. released *Torch and Twang*, her farewell to country, in 1989. Influenced by the big skies of Alberta and Montana, k.d. and Ben Mink set off on a "mood searching drive" through the badlands. After the hiatus of *Shadowland* k.d. elected to return to Mink and The Reclines. Her relationship with Mink started back in '85, when she went to play at Expo in Japan. When they met, Mink showed her his fiddle which he kept filled with little treasures. Hidden away inside the body was a world of toy soldiers, bathing beauties, plastic farm animals, a bagel, and a flag from Nashville. k.d. asked if Mink had any songs. That was the beginning.

Ben's a Canadian, from Toronto, and couldn't come from a world more removed from k.d.'s His parents, Hasidic Jews, bear tattoos from the Nazi Holocaust. He is the grandson of a trained Hasidic singer. Raised in the big city, Mink represents the cosmopolitan antipole to k.d.'s rural roots. The two feed off each other's energy and share equally meticulous standards of composition. When they collaborate, k.d. pens the lyrics and develops the idea, Mink then tosses out a riff, a drum pattern, setting the mood. Together they sort out what's good. Mink works from a technical standpoint, constantly recording and cataloguing sounds on tape. k.d. works from instinct, in a primal arena. She values Mink because "he manages to capture my little ideas".

It's no coincidence that both *Absolute Torch and Twang* and *Ingénue* represent k.d.'s finest work. Both albums reflect k.d.'s decision to let go, collaborating intensely with Mink without censoring herself or trying to give people what she thinks they want. *Lariat* and *Shadowland* succumbed to a desire to hit it big on the first time. On *Torch and Twang* k.d. stepped back from

the world of Nashville and returned to her roots, the big skies of the West. She cut the album in Vancouver and the result was beautiful. *Torch and Twang* pushes to the edges of country, bringing in influences from jazz, to flamenco, and even Eastern European traditions. She and Mink, from such seemingly different worlds, actually have a great deal in common, not the least of which is k.d.'s recent discovery that she too is Jewish.

Absolute Torch and Twang sold over 1.1 million copies, beating out *Shadowland*. k.d. and Mink wrote most of the tracks together. They produced the album along with Greg Penny, a relatively unknown country producer who let k.d. do it her way. As usual, country radio avoided it. The tracks were compared to a cattle prod up country's ass. It's a lot more than that. It's the first time k.d. looked into herself and revealed that much.

Wallflower Waltz hints at homosexuality, as the singer waits for someone to ask her to dance. *Trail of Broken Hearts* features k.d. wandering through the big blue sky of Alberta, looking "handsome" (as her mom called her) singing with all the rawness of open land. *Big Boned Gal* tells the tale of a "big boned gal from southern Alberta" who'd hold her audience in a trance, she's a "natural". That was it. Country lost it. k.d.'s androgyny no longer parodied, it broke down barriers, coming into full sensual bloom. k.d. waved goodbye to the thought of ever fitting in and hurtled down her own path. Finally free.

You can feel the relief in all of k.d.'s work from 1989 on. It just keeps exploding, getting better and better. This is the k.d. poised to go into every medium. 1990 pretty much smashed any remaining barriers to bits. The PETA "stink" fumed its way across the land, severing k.d. forever from Nashville. She walked away, heading for Cole Porter, *Salmonberries*, and the coming out that would grace *Ingénue's* release. Even a hint of disco lurked around the corner.

torchy twangy roots

beyond androgyny

"Androgyny to me is making your sexuality available, through your art, to everyone... using both the power of male and female."

"I have been called 'sir' so many times in my life and will always be."

The clues about k.d.'s sexuality became progressively more and more overt as her profile became higher and higher. It began with her video for *So In Love,* her track on the *Red Hot & Blue* compilation that raised money for AIDS research. Then *Salmonberries* came out in 1991. k.d. came out a year later. Both songs deal with love between women.

Percy Adlon (*Baghdad Café*) directed k.d.'s *So In Love* video, an outstanding collaboration. k.d. washes her dead lover's clothes, hanging a woman's nightshirt to dry, singing "I'm yours 'til I die". Shots of an I.V. bag dripping, a plastic invalid's seat in the empty shower, water dripping, k.d. wanders alone amidst the traces of her lover's passing. Adelon went on to write a role for k.d., putting her in his film, *Salmonberries*.

k.d. believes she was always out, she never denied her sexuality, she just didn't shout it from the rooftops. *Salmonberries* walks that quintessential androgynous line, then goes right over it. k.d., looking every inch a man (she plays an oilfield worker), takes off her parka and reveals her curving, womanly body. She stands, naked, facing the stunned librarian. The object of seduction. This tale of love is a tale k.d. understands. Their aborted lovemaking scene fits k.d.'s experience of her first lesbian encounter, "I know that one of the first times I ever had someone come on to me, I sort of went, 'No.' Then, 'Well...'." *Salmonberries* hasn't found an American distributor yet. Maybe that's why so many found her coming out to *The Advocate* surprising. k.d. came out to her mom at 17, no problem there; her older sister Keltie is gay too. There's an assumption that coming out is tortuous and painful, especially in a small town, yet for k.d. it wasn't. Her background, on the surface, fits the stereotypes of an isolated girl in an isolated town, yet that's not the case.

the mind of love revealed

*"**Ingénue** means unworldly, naïve, artless. An unworldly artless woman played by an actress."*

*"**Ingénue** is based on my experiences of falling in love, and it's the most personally revealing record I've ever made...naked and real... if I was toothpaste and you squeezed me, you'd get **Ingénue**."*

"During this time I fell in love for the first time and it's an overwhelming experience. The theme of the record is obviously love but it's getting to the point of love where you realise you have no control and you can't manipulate it. And that love is about pain as well as pleasure. It's about relinquishing control to love. I experienced a period where I thought, Wow, I'm not going to conquer the world like I thought I was going to. I don't have it on a string. It's a period of great awakening and realising that this is the way you are and you have to live with it."

"You know, I've never seen a blue heron with another blue heron, have you? You know blue herons? They're always alone. But they must mate."

For k.d. the passion for country music "just died". And love filled the void, an impossible private love fated to end. That severing spawned *Ingénue*. k.d. fell in love with a married woman and the impossibility of the relationship broke open a well of feeling, it was "an unattainable love... a feeling for somebody that I can't have right now". *Ingénue* became the catharsis; k.d. freed herself to explore and explain the "mind of love". She calls the album "emotional puberty". While the songs focus on unrequited love, she believes it has a positive undertone: she rediscovered her first lover – music.

k.d. faced an inner challenge, she'd never revealed this much of herself in public. She clings to her privacy, and that clinging stood as a potential barrier to letting herself escape into the music. When she and Mink got together to begin writing *Ingénue*, k.d. discovered she was totally blocked up, unable to sing. Until now, k.d. admits, her singing had an element of method acting: she'd get into character and from that place she'd sing (the Patsy Cline period). That worked fine when her music remained somewhat removed from her private experiences. But the point of this album was to move to another place with the music.

k.d. feared that she couldn't do it, couldn't let go of her self-control and let herself be drawn by the music, a journey similar to the cycle she experienced with her lover: letting go, then getting hurt, then left to pick up the pieces. She had to turn inward, and for years she'd experienced singing as ascending outward into character,

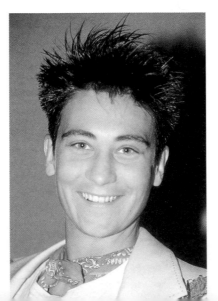

being the entertainer. When she broke through that barrier, she completed the cycle of unrequited love, "catharsis", as she calls it. Getting there took time – and a trip to Europe.

k.d. went to Europe to promote *Salmonberries*. She needed a break from the studio in Vancouver. She'd started to track the vocals and they weren't turning out properly. k.d. was "still singing from the writer's perspective. I thought I'd lost it. I thought that my voice was taken away from me." When she got to Europe she wandered alone, walking through the streets of Paris, Berlin, Stockholm, and Zurich. She discovered she'd been singing from the wrong place. Returning to Vancouver, she "got it". On her first day back, k.d. and Mink put down vocals for four tracks.

When k.d. sings, she's brutally honest. She'll sing in front of a mirror, and if the voice isn't authentic, she won't accept it. For years she'd been singing as the "yahoo k.d.", according to Mink. When she got back, she'd developed a new language, aspiring to use "the power of subtlety". Finally, for Mink, "the words and phrases began to make sense". k.d. and Mink locked themselves up in a room, "like two monkeys", and set up a trapeze of instruments. In their collaborative style, they worked through everything, intellectualising later. It took them fifteen minutes to come up with *Miss Chatelaine*. The album took a week and a half to write, only later did they agonise over the arrangements and lyrics. During that time, k.d. stopped listening to country music completely. She took in Karen Carpenter, Nat King Cole, Joni Mitchell, Carmen McCrae, Peggy Lee, Julie London, and Yma Sumac. Kurt Weill and Berthold Brecht appeared as influences (*Constant Craving* is inspired in part by *Waiting for Godot*). They were moving deep into new territory.

Released in April 1992, *Ingénue* went platinum, selling over two million copies to date. The sound is totally different from her previous work, but the crooner in k.d. remained. She took the torch and skipped the twang. The album owes little to country. The tracks are uniformly moody and introspective, apart from *Miss Chatelaine* and *Constant Craving*, reflecting a combination of sultry big-band influences and Parisian cabaret (k.d. calls the mix "postnuclear cabaret," whatever that is). Popping up here and there are traces of flamenco, Middle Eastern music, even some Yiddish klezmer music manages to pay a visit. This time, DJs were friendly. The album got airplay and recognition with a Grammy for Best Female Vocalist. A month after its release, k.d. met with Brendan Lemon and told *The Advocate* she wanted to defuse all the speculation about her sexuality. When k.d. came out she wondered what the impact might be on her career, would people stop buying her albums? She looked at Freddie Mercury, whom she called "gay as a daffodil", as a musical example of public acceptance.

coming out to the world

"I'm not comfortable with it... my body is very womanly; I think that's why. Or maybe because it's so overused in the entertainment industry. Maybe it's a deep rebellion. I don't understand my own feminine power yet, in terms of my body. I don't know how to use femininity as a powerful tool. I use my sexuality, but I eliminate the gender from it. I think it's important for people to come out because it's broadening the acceptability walls. But I always thought I was out. I presented myself as myself. I didn't try to dispel lesbian rumours. I sang songs like *Bopalina* which was about my girlfriend."

"I don't know why I'm gay. I find women more enticing, both emotionally and sexually."

"If you want to know the truth, I am so much a woman. I am as womanly as I could be. With my lovers, I'm completely a woman... I think the male thing is just a way of surviving – outside. Inside I'm completely a woman."

"Androgyny is important in my life because I can deal with people on a human, not a sexual level; it's important on stage, because both men and women are attracted to me."

"(Coming out) was totally positive, totally positive. Like an emotional veil had been taken away. The really, really big thing I experienced this year was the intimacy between me and the audience... I feel comfortable knowing that they came there knowing."

Gender-bending images are not new to popular culture. k.d.lang stands in a place pioneered by Little Richard and Elvis, taken up by David Bowie, Michael Jackson, Annie Lennox, Madonna, and Prince. All these performers spoke to a special place, the delicate cusp between male and female: acts that highlight cultural anxiety about change. They seek to blow apart natural instincts which see the world in terms of simple opposites: male/female, rich/poor, black/white, gay/straight. This is the "middle of the road" k.d. talked about, a place where anything is possible, where art brings new meaning to standard assumptions. It's a place where many find it uncomfortable to dwell and performers who approach face the flack that arises from the fear of change. What they represent poses a threat. Their very being questions what is "normal", and ultimately redefines it. Thirty years ago a lesbian would have found it impossible to ever achieve the kind of acceptance that k.d. has enjoyed. Even ten years ago, David Bowie's carefully orchestrated 1971 publicity campaign aside, the assumption was that to declare one's homosexuality guaranteed public hostility and a fast path to oblivion.

It is mainly due to the actions of artists who preceded k.d. that only one of her CDs was ever returned to Sire

Records because of her admission (yet thousands were returned during the PETA controversy). There comes a moment when years of change come together in a single symbolic climax and the world suddenly takes stock of a seamless progressive shift in values, and thus ushers in a new era. k.d.'s coming out was such a moment, and the response reflects the profound change contemporary culture's experienced since the days of Little Richard.

Today white kids in the suburbs buy records by a six-foot-seven drag queen named RuPaul and no one thinks twice about it. In the spring of '93, a million gay men and women marched on Washington D.C. shouting that they will be heard. An out lesbian holds a cabinet level post. The U.S. military faces a President who wants them to admit avowed homosexuals to their ranks. Something is going on here that we have yet to make sense of, and it's no coincidence that it all exploded in the same year: it is the crest of the wave that started building years ago. Its origins, long passed, paved the way for Kathy Dawn.

k.d. is not simply a singer, she is a campaigner for a new aesthetic that's come into its own. In her she carries the responsibility and the gift to make sense of the changes contemporary culture's witnessed. When she looks to the future, she acknowledges her special place and the privilege her art gives her. She considers herself first and foremost an artist, and all the other categories (lesbian-vegetarian-feminist-pantheist) come second. She wishes to transcend these categories, presenting what connects humanity, not what separates it.

always just beginning

"I think I've been really honest and open about the fact that I'm not married to any one genre and I've changed in the past and I may change again. For all I know the next record might be heavy metal".

"Artists are put in a trap. They attain success, and there's pressure to do it again, so they basically just write the same record again... One of my goals is to keep dissatisfied. I am very conscious of not checking out and just going through the motions."

"To be a star, to me, was saying I want to sing, I want to be an artist. I've achieved some of my material goals... My goal now is to stay creative and have a career of longevity and history. I want to feel the muses, to sustain the reciprocity between me and what makes me an artist. I intend to stay an artist of integrity with brave force, to stay moving in a straight line at a direct pace. It boils down to that."

"The rockabilly rebel's still there, she's still inside, and it's like at the same time you're driving forward, you're waving goodbye, kinda unwillingly, at the beauty of being wacky and immature...I'm changing all the time. Someone remarked to me the other day, 'Champions adjust.' So anything could happen. I could meet a Tibetan monk and shave my head bald."

In the fall of '93, the film *Even Cowgirls Get the Blues* (directed by Gus Van Sant) comes to theatres in America. k.d. and Mink worked on the score. It's a fusion of early '70s styles (the film takes place in 1973), from polka, jazz-fusion, country waltzes, and Sly and the Family Stone boogie tunes. As soon as the tracks are completed, k.d. goes back into the studio to produce her next album. There's a lot of pressure to repeat *Ingénue*, but she refuses to write the same record again. Apparently the pressure is on to make money while she's hot, and k.d. insists on not "producing for success". It's a trap she fell into with both *Angel With A Lariat* and *Shadowland*. She's also considering film roles, this time ones that explore her feminine side. She's interested in doing a straight feminine character.

Not content to stand with what she knows, she keeps looking for inspiration in undiscovered places, places she may not even be aware of yet.

SINGLES...

Rose Garden

High Time For A Detour

Sire W 8465 (7") Unreleased

Sugar Moon/Honkey Tonk Medley

Sire W 7841 (7") June, 1988

Sugar Moon/Honkey Tonk Medley/I'm Down To My Last Cigarette

Sire W 7481T (12") June, 1988

Our Day Will Come/Three Cigarettes In An Ashtray (live)

Sire W 7697 (7") November, 1988

Our Day Will Come/Three Cigarettes In An Ashtray (live) Johnny Get Angry (live)

Sire W 7697T (12") November, 1988

Constant Craving (edit)

Barefoot (Rhythmic Version)

Sire W 0100 (7") April 1992

Constant Craving (edit)

Barefoot (Rhythmic Version)

Sire W 0100C (cassette) April 1992

Constant Craving (edit)

Barefoot (Rhythmic Version)

Season Of Hollow Soul

Sire W 0100T (12") April 1992

Constant Craving (edit)

Barefoot (Rhythmic Version)/Season Of Hollow Soul

Sire W 0100CD (CD) April 1992

Crying (with Roy Orbison)

Falling (Orbison)

Virgin America VUS63 (7") August 1992

Crying (with Roy Orbison)

Falling (Orbison)

Virgin America VUSC63 (cassette) August 1992

Crying (with Roy Orbison)

Falling (Orbison)

Oh Pretty Woman (edit) (Orbison)

She's A Mystery To Me (Orbison)

Virgin America VUSCD63 (CD) August 1992

Crying (with Roy Orbison)

Falling (Orbison)/Only The Lonely (Orbison)/It's Over (Orbison)

Virgin America VUSCX63 (CD) August 1992

Miss Chatelaine (single version)

Miss Chatelaine (St Tropez edit)

Sire W 0135 (7") September 1992

Miss Chatelaine (single version)

Miss Chatelaine (St Tropez edit)

Sire W 0135C (cassette) September 1992

Miss Chatelaine (St Tropez mix)

Miss Chatelaine (St Tropez edit)

Miss Chatelaine

Miss Chatelaine(single edit)

Sire W 0135TW (12" with poster) September 1992

Miss Chatelaine (single version)

Miss Chatelaine (St Tropez edit)

Wash Me Clean/The Mind Of Love

Sire W 0135CD (CD) September 1992

Constant Craving (edit)

Miss Chatelaine

Sire W 0157 (7") February 1993

Constant Craving (edit)

Miss Chatelaine

Sire W 0157C (cassette) February 1993

Constant Craving (edit)

Wash Me Clean/The Mind Of Love

Sire W 0157CD (CD) February 1993

Constant Craving (edit)

Miss Chatelaine

Big Boned Gal/Outside Myself

Sire W 0157CDX (CD) February 1993

UK ALBUMS...

Angel With A Lariat

Turn Me Around/High Time For A Detour/Diet Of Strange Places/Got The Bull By The Horns/Watch Your Step Polka/Rose Garden/Turn Into My Wave/Angel With A Lariat
Sire K 925441 February 1987

Shadowland

Western Stars/Lock, Stock And Teardrops/Sugar Moon/ I Wish I Didn't Love You So/ Once Again Around The Dance Floor/Black Coffee/Shadowland/Don't Let the Stars Get In Your Eyes/Tears Don't Care Who Cry Them/I'm Down To My Last Cigarette/Too Busy Being Blue/Honky Tony Angels (medley)
Sire WX 171 April 1988

Absolute Torch And Twang

Luck In My Eyes/Trail Of Broken Hearts/Didn't I/Full Moon Full Of Love/Big Big Love/Walkin' In And Out Of Your Arms/Three Days/Big Boned Gal/Wallflower Waltz/Pullin' Back The Reins
Sire WX 259 May 1989

Ingénue

Save Me/The Mind Of Love/Miss Chatelaine/Wash Me Clean/So It Shall Be/Still Thrives This Love/Season Of Hollow Soul/Outside Myself/Tears Of Love's Recall/ Constant Craving
Sire 7599-26840-1

CANADIAN SINGLE...

Damned Old Dog

Friday Dance Promenade

Bumstead (500 copies) 1983

CANADIAN ALBUM...

A Truly Western Experience

Bumstead 1984